T0274184

THE
WASTEWATER
GARDENER

THE WASTEWATER GARDENER

Preserving the Planet

One Flush at a Time!

MARK NELSON

FOREWORD BY TONY JUNIPER

SYNERGETICPRESS
expanding human knowledge

SANTA FE AND LONDON

© Copyright 2014 by Mark Nelson. All rights reserved.

No part of this publication may be reproduced, stored in any retrieval system, or transmitted, in any form or by any means, electronic, mechanical, photo-copying, recording, or otherwise without the prior permission of the publisher, except for the quotation of brief passages in reviews.

Published by Synergetic Press
1 Bluebird Court, Santa Fe, NM 87508

The Wastewater Gardener © Mark Nelson, 2014
Foreword © Tony Juniper, 2014

Library of Congress Cataloging-in-Publication Data

Nelson, Mark, 1947-

   The wastewater gardener : preserving the planet one flush at a time! /
   Mark Nelson ; foreword by Tony Juniper. -- First edition.
   pages cm
   Other title: Preserving the planet one flush at a time!
   Includes bibliographical references and index.

   ISBN 978-0-907791-52-2 (hardcover) -- ISBN 978-0-907791-51-5 (pbk.)
   1. Sewage irrigation. 2. Water reuse. 3. Sewage--Environmental aspects.
   4. Gardening--Environmental aspects. 5. Constructed wetlands. I. Title.
   II. Title: Preserving the planet one flush at a time!

TD760.N35 2014   2014009840
628.1'68--dc23

Cover concept by David Rogers.
Cover art, illustrations and map by Jeff Drew
Cover photo of author by Stephen Muller
Book design by Ann Lowe
Editors: Linda Sperling with Hugh Elliot

Printed by Friesens Printing, Canada
The text of this book was printed on 60# Husky Offset White 100% recycled post consumer waste paper.
Typeface: Adobe Garamond Pro and Helvetica Neue Condensed

This book is dedicated to Susannah Garrett, mi Fortuna,
who makes me appreciate my *suerte* every day.

Table of Contents

Foreword

THERE ARE SEVERAL MODERN SYMBOLS of ecological crisis. Gas-guzzling vehicles, airliners, coal-fired power stations and landfill sites are among them. While few people would add flush toilets to the list, there is increasingly good reason to see why that might be the case.

Freshwater is one of those day-to-day necessities that many of us have become used to taking utterly for granted. Delivered clean and safe through pipes to homes and offices, it can seem like an endless resource that will always be easily available. Reality is somewhat different, however, for in a planetary sense freshwater is not as abundant as it can sometimes seem to be, far from it in fact.

Of the 1.4 billion or so cubic kilometers of water that we have on Earth, nearly all of it is in the oceans and therefore salty. About 97.5 percent of it is in that form. Of the remaining 2.5 percent that is fresh, some 60 percent is trapped in ice caps and glaciers with 30 percent more in groundwater, and therefore for the most part not immediately available to us for farming, domestic supply and industry. The tiny remaining amount that is generally renewable and in rivers, lakes, ponds and clouds, is subject to increasing demand from more people living in larger, growing and ever more demanding economies. Not only is that little slither of freshwater under mounting pressure, its local availability is also subject to growing volatility because of changes taking place in the Earth's climate.

We go to great lengths to ensure that societies have enough freshwater, and each year spend billions of dollars on extracting it from the environment, putting it in reservoirs, cleaning it and then piping it to where it's needed. The precious resource to which we go to such lengths to supply is every day then turned into "wastewater" of different kinds, including by the simple act of flushing lavatories.

This convenience was of course invented for good reason and the sanitary engineers who during the nineteenth and twentieth centuries contributed to the rise of the modern toilet can correctly be seen as among those who helped build the healthy lifestyles so many of us enjoy today. They helped to reduce the spread of infectious disease and brought to the mainstream devices that would increase people's lifespans.

At the same time though, they introduced technologies that would require some ten thousand tons of water to remove and treat each ton of human excrement and lead each one of us with a flushing toilet to use in the order of ten thousand gallons of water a year to send our waste away.

Increasingly though, we are finding that in our shrinking world there is no "away." In the end all of our waste ends up somewhere, often contributing to different environmental problems when it arrives at its final destination. Considering the dual trends of rising demand for freshwater and the increasing amount of nutrients leaving our toilets building up in the environment, those problems will only become more complex. Fortunately though there are solutions, and in this marvelous book Mark Nelson shares his decades' long experience to explain what they are.

His basic conclusion is that by copying nature, and in particular how wetlands work, cost effective, healthy and sustainable alternatives to our wasteful sanitation systems can be put in place, not only defending us from the effects of water scarcity, but also protecting ecosystems from impacts that can be caused by too much of our bodily wastes being released into them. He takes readers on an inspiring journey, one that is rendered all the more impactful by his stories of doing it for real.

From Mexico to Western Australia and from Bali to Arizona, Nelson has designed and installed systems that have provided local sanitation solutions while bringing global benefit and his stories show how he's done it with humor, humility and humanity.

Painting a picture of how soils, plants and water might be harnessed in ways that meet our needs without pipes, sewage works and pollution, Nelson inspires us to see opportunities that are not only practical, but if designed well also beautiful. Channels of water flowing between willows, water hyacinths, lilies, rushes and cattails, with the cleaning function of roots and microbes harnessed in the purification of water, could almost not be more different from high-tech centralized wastewater treatment facilities.

The construction of beauty is, however, a very practical response and one predicated on the understanding that the excrement we do so much to get rid of is essential to life. All living things require nutrients to exist, grow and reproduce and the material we eject from our bodies and into sewers is a massive source of those life-giving materials. That we have lost sight of this basic reality and strive so hard to place ourselves outside nature is one further manifestation of the industrialized mindset that any rational reading of our direction of travel would suggest we need to change.

In common with others who present the kinds of durable solutions that are needed to deal with the multiple challenges that

confront us, Nelson describes how it is not only different treatment systems that are required, but also a shift in this unrealistic
mindset. Especially through how we might see things differently
by taking more of a system-based approach, compared with more
reductionist ways of thinking that seek to deal with each problem
at a time and in isolation from related ones.

Perhaps through the pages of this important book the flush
toilet might one day become not only a symbol of unsustainable
development, but also of the hazard that can accompany "solutions"
that meet one challenge while not considering others. For while the
flush toilet has been successful in protecting public health, it will,
because of, among other things, progressive water scarcity, nutrient
depletion and environmental impact, not be a solution that can
lend itself to upwards of nine billion people.

This is why *The Wastewater Gardener* is an essential read for anyone
with an interest in sustainable development, water, urbanization,
sanitation and public health. For while it might seem that the
only alternative to the open sewers that characterized so much of
our past, and in too many places the present, is the installation of
flushing toilets and building sewage works, there is in many cases a
better way, one that brings not only health and sustainability, but
also beauty. Infusing it all is the realization that nature does not do
waste, and if we wish to endure, then neither should we.

—Tony Juniper
Environmentalist, Author, Fellow with the
University of Cambridge Institute for Sustainability
Leadership and Advisor to The Prince of Wales
International Sustainability Unit

Preface

"Anyone starting out from scratch to plan a civilization would
hardly have designed such a monster as our collective sewage system.
Its existence gives additional point to the sometimes asked question,
Is there any evidence of intelligent life on the planet Earth?"
—G.R. Stewart, author, *Earth Abides*, 1949.

THIS BOOK HAD ITS GENESIS the first time I tipped over an outhouse and shoveled the steaming contents into a wheelbarrow headed for the humanure compost heap. I was a city kid; I didn't know the stuff was taboo.

When I was selected to be a "biospherian," that is, a crew member for the first two year closure experiment of Biosphere 2, it may have been destiny that one of my responsibilities was managing our "marsh recycling system" for all the wastewater. Later on, when I'd fallen in love with wetlands, both natural and constructed, I decided to make myself useful by confronting sewage problems around the world. That commitment led me from one improbable adventure after another to a veritable Wonderland of strange goings on. Fortunately, I held to my inner yoga and can testify that by keeping your optimism and belief you *can* make a difference, and never lose your sense of humor!

It occurred to me these "adventures in the shit trade" have more than purely humorous anecdotal value. I got to see what is hidden for good reason from most people, although it sometimes took persistence and detective work to find out what was really happening. I feel a responsibility to share with a greater audience what I have seen and learned. As ecologists say, everything is connected to everything, and how we manage and mismanage our shit, is a crucial part of the global challenge of our times.

Conventional industrial-style agriculture doesn't use animal manure—we turn our farms into monocultures, raise our animals in factory settings, using lots of chemical fertilizers which are expensive, releasing greenhouse gases and nutrients run off from our farms in great quantities polluting our waters and oceans. In the West, we centralize sewage treatment—sending all sewage nutrients into our rivers and oceans, instead of back to our farms or green spaces. Rather than irrigate using graywater, we use precious, high quality potable water. In poorer countries, there is virtually no effective sewage treatment at all and widespread contamination of drinking water leads to disease, death and further impoverishment.

We all know the Hans Christian Anderson tale, *The Emperor's New Clothes*, in which only a child is honest enough to speak aloud of the emperor's delusion. This book is the global black comedy which unfolds when the little boy opens his eyes. I hope it changes the way you think about at least one of the so-called "little" things we do in life.

I gratefully acknowledge the many colleagues and participants whose efforts led to the creation of the work contained in this book. These include my Institute of Ecotechnics friends and teachers: in particular John Allen, a rare man who combines vision and common sense, and also Marie Harding, William Dempster, the late Robyn Tredwell, Margaret Augustine, Chili Hawes, Gerard Houghton, Kathelin Gray and Deborah Parrish Snyder.

My Wastewater Gardens International intrepid network of designers and regional representatives: especially Florence Cattin who made our Algeria work possible and brought the technology to new countries like Spain, Portugal, Morocco, the Maldives and some parts of Indonesia; also Davide Tocchetto, Gonzalo Arcila, Ingrid Datica, Andrzej Czech, I Gede Sugiartha, Andrew Hemsley, Malini Rajendran and Lucien Chung.

I deeply appreciate the leadership and support of Abigail Alling and Mark Van Thillo of the Planetary Coral Reef Foundation through which we built the early Wastewater Garden systems in Mexico, West Australia, Bali and Sulawesi, Indonesia. The Foundation contributed significantly to the creation of Wastewater Gardens International and its network. And Emerald Starr whose passion for the environment helped make possible the many lush Wastewater Gardens flourishing in Bali and around the Bunaken Marine Reserve in Sulawesi. And our team in Mexico which included Gonzalo Arcila, Ingrid Datica, Reka Komaromi and Klaus Eiberle whose dedication and hard work led to building over 50 systems along the Yucatan coastline. At the University of Florida, the late H.T. Odum, Mark T. Brown, K. Ramesh Reddy, Clay Montogue, and Daniel Spangler taught me how to think in systems; working at their Center for Wetlands sparked the creation of new approaches to ecological engineering.

Wastewater Garden projects have received the kind support of many key people and organizations including Petra Schneider and IDEP Foundation, Indonesia, Seacology Foundation, the Livingry Foundation, the Sendzimir Foundation, the Puerto Rican Department of Natural Resources, Maurice Levy of Earth and Water, Portugal, Sheikh Mohamed Laïd Tidjani, head of the Tidjani Sufi order in Temacine, Algeria, the Algerian artist Rachid Koraïchi and the Association Shams, Abdelkader Belkacemi and Mohamed Sadaoui (Faouzi) from the Algerian Ministry of Water Resources, the Belgian Technical Corporation, Lamine Hafouda from the INRA station at Touggourt, Algeria, Juan José Salas Rodríguez from the Centro de las Nuevas Tecnologías del Agua of Andalucia and the PECC, Ben Brown of the

Mangrove Action Project, the World Wildlife Fund, David Carpenter at the Aboriginal Housing Division of the West Australian Department of Housing and the Australian Community Water Grants.

I wish to thank Sally Silverstone and Thrity Vakil, previous and present managers of Las Casas de La Selva who facilitated building our demonstration system in the heart of the sustainable tropical forestry project in Puerto Rico. And my colleagues in Australia: Robyn Tredwell who selected the native plants and *bush tucker* (food) plants for our systems on the Aboriginal communities in the Kimberley and who skillfully navigated our projects through unchartered political and cultural waters; and her team from Birdwood Downs Company of Greg Hay, Hans Leenaarts, Brad Riley and Eddie Zuna.

The visionary leaders at Nature Iraq, Azzam Alwash, Jassim Al-Asadi and Ammar Zakri, along with the gifted artist, Meridel Rubenstein, are helping realize our ground-breaking project, *Eden in Iraq*, with the Marsh Arabs in southern Iraq. Our design team also includes Prof. Sander van der Leeuw of Arizona State University and Prof. Peer Satikh of Nanyang Technological University (NTU). And thanks to NTU for a grant making possible our site visits and design work.

I am grateful to the publisher, Deborah Parrish Snyder, and Omar Fayed, president of Synergetic Press for seeing the importance of this book; and the Synergetic Press team: David Rogers, Debbie McFarland, Mitch Mignano, Melissa Guthrie and Graciela Ruiz for making this such a fun book. The surgical work of Hugh Elliot and Linda Sperling, my gifted and dedicated editors, is reflected throughout the book.

There's a saying in my Institute: "Beauty, Discipline, Honor and Friendship." May these be watchwords as we strive to remake the world a little closer to our heart's desire.

—Mark Nelson

1

A Brief History of How We Got Into This Mess

> *"He can't tell the shit from the shinola."*
> —American folk saying for incorrigible stupidity.

THERE'S A FOUR LETTER WORD that still awaits liberation: it is considered to be far nastier than sex, which is used to sell the global economy's products, to entertain and enthrall people and to power our gossip, soap operas and talk radio. But *this* word is never brought up in polite society, certainly never at dinner table conversations, and is equally repugnant to Left and Right. Yet it's everywhere, certainly everywhere there are people.

We employ endless euphemisms to avoid having even to say it or to think about its use and abuse: from baby talk "caca" and George W. Bush's "deep doo-doo," to hands up in school to request priority permission to take care of "number two." It's the New Agey organic "humanure," the dryly academic "excrement" or "human solid waste product."

It's also the somewhat archaic "night soil" and in the jargon of sewage professionals and recyclers, it's "black water," as opposed to "graywater" (all the other "wastewater" from the laundry, kitchens, sinks and showers). People go to great lengths to avoid seeing, smelling, touching, thinking about or dealing with it.

1

Of course, I'm talking about "shit," a word for which many languages have negative connotations, as in "*merde*" (French) or "*Scheiss*" (German).

Sigmund Freud defined the progress of sexual development in human psychology from oral, to anal, to genital stages. Contrast the very young child's sensuous pleasure in the process of defecation with the widespread anal-retentive behavior seen in many adults for whom the very mention of the word "shit" produces consternation.

But what if I told you that in medieval Europe, this same substance was valued as a medicine and even used as a cosmetic for preserving the youthful appearance of a woman's face?[1] As we will later discuss in more detail, throughout the world the use of human as well as animal manure fertilizer endowed it with great value. It was regarded as a resource, not as waste to be disposed, at least in Asia until the beginning of the twentieth century. In Japan and China, "shit" is still never used as a negative word. During the Korean War US soldiers were astonished to see Korean farmers entice travelers to use their outhouse toilets so they could collect their excrement.

Ponder a popular exclamation to express wonder. The ancient Manicheans viewed the entire cosmos as a turd of God, yea verily, it is the *Holy Shit*!

Facts and figures:

An adult human produces about one half to one pound of waste per day. With more than seven billion people on the planet, that comes to at least three and a half billion pounds or 1.75 million tons per day; over 600 million tons, per year!

When there were only a few million humans scattered in small groups around the world, the danger of pollution from human waste was fairly small. Most cultures have an instinctive aversion to

[1] La Porte, Dominique: *History of Shit*, trans. Nadia Benabid and Rodolphe el-Khoury, MIT Press, Cambridge MA, 2000, Chapter 5: Make-up, pp. 96-112.

Shanghai: circa 1911. A flotilla of boats used by contractors who buy "night soil" and transport it upriver to sell to farmers for composting and application to their fields. From: F.H. King's *Farmers of Forty Centuries*.

fouling the nest. People used to go out to the woods or meadows to defecate, depositing their humanure on the soil where microbes can easily decompose it and make its nutrients available for plants. Human feces can cause a multitude of diseases, including major killers endemic in the developing world: diarrhea, cholera and typhoid. Contamination and the spread of disease are found even in cultures renowned for their ecological sustainability. In the rice paddy networks on the island of Bali and elsewhere in Southeast Asia, infant mortality rates remain high partly because drinking water is polluted by human shit.

Many cultures used to acknowledge the value of humanure as a fertilizer and developed composting practices where it was mixed with plant and vegetable waste to make topsoil. They handled the substance carefully, seeing it as a natural resource. Among the traditional cultures of China, Japan and Korea, the humanure that was produced in cities neither contaminated drinking water supplies, nor cost a city a fortune to dispose of or treat. These "farmers of forty centuries" understood the importance of returning the nutrients contained in humanure back to the farms. "Night soil" was collected by contractors who paid for the opportunity. The material was loaded into boats and sent up river to be sold to farmers who then used it for making compost or directly fertilizing their fields and crops. It was simple: a farmer grew food, shipped it to the city whose waste products were then collected and returned to the countryside. These cultures were able thus to maintain soil fertility, century after century. For the cities, the shit was a source of revenue, not a problem requiring the expense of modern sewage treatment facilities. The idea of throwing away a valuable resource and causing pollution was inconceivable.

Indoor Plumbing

As human population increased and became more urbanized, traditional collection and reuse methods became more difficult to implement. In the West, an understanding of the value of human waste as fertilizer was superseded by the convenience of using chemical fertilizers. Farm animals like draft horses and oxen were being replaced by fossil fuel-driven tractors, so a huge source of animal fertilizer and compostable material was now gone. Farmers used to be generalists, raising animals and crops. Modern farmers tend to be specialists. Today's farms are either meat factories, where animals are raised in automated facilities and the animal waste is a problem that ends up polluting rivers and groundwater, or they are crop specialists, relying on chemical fertilizers. An unintended consequence of crop specialization is runoff of chemical fertilizers into the environment. Fertilizer runoff and sewage are the leading causes of water pollution on the planet. Chemical fertilizers, though so convenient and easy to apply, are both expensive and costly to the environment. Their production and long distance transport significantly add carbon dioxide and contribute to climate change. As much as half the chemical nitrogen applied to crops and 20% of the phosphorus, washes off into groundwater, rivers, lakes and ultimately into the oceans.

There are those who see indoor plumbing as the mark of a civilized society and the solution to the epidemics that periodically broke out prior to the nineteenth century. Instead of having to venture out, especially on cold nights, to an outdoor toilet or privy, you could now conveniently defecate in the comfort of your own home. And to remove the waste, just add fresh, potable water! One benefit: no unpleasant odor when workers came to remove the contents of the outhouse. So now we have a centralized system. Send the humanure to a municipal sewage treatment plant, where trained specialists detoxify the shit and make it fit for final disposal—note: disposal, not usage.

In less densely populated urban areas and in the countryside indoor plumbing is generally attached to a septic tank and leach field disposal system. The sewage solids are separated in the septic tank, and the remaining wastewater is percolated through perforated pipes laid in trenches of gravel in the hope it will be cleaned before it reaches groundwater. Solids (sludge) that are pumped out from the septic tank are taken to the centralized sewage treatment facility. Except, of course, in poorer countries where the sewage hauler might head for the nearest jungle, mangrove swamp or deserted lot to dump the load. Dumping costs less: less fuel expenditure and no fees need be paid to the sewage plant.

Evidence of indoor plumbing and the use of water to carry away human effluent can be traced back to examples in India, Scotland, and the famous Cloaca Maxima (Greatest Sewage System) of ancient Rome. Its modern reinvention was centered in northern Europe, especially England, starting in the late fifteenth century. Indoor plumbing had vastly improved by the late nineteenth century and rapidly caught on in the great cities. Urbanites had the astonishing modern convenience of having their water piped, rather than hauled, into their homes. But don't forget "the law of unintended

A typical arrangement for getting wastewater from homes and industry to the centralized treatment plant with underground pipes and pumping stations. Combined sewage overflow indicates the use of this system for also carrying rainwater, which often results in overloading during heavy rains and diversion directly to the river/ocean.

At every step of the way, this treatment and disposal approach requires money, machinery and energy, and leads to pollution wherever the wastewater and sludge is sent.

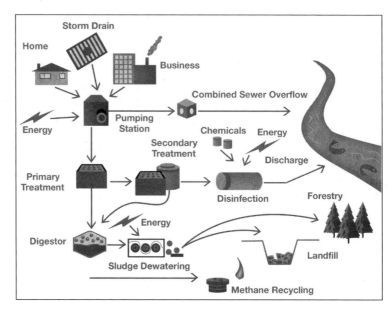

consequences" and what then came about due the wonderful convenience of indoor plumbing.

While some praise indoor plumbing and the flush toilet as sterling achievements, for others, it is the height of insanity to use drinking water to dispose of human waste and then wash it away into large bodies of water, spreading the potential for pollution of all Earth's water bodies.

When per capita fresh water usage in towns and cities was low because water had to be hauled into houses, waste was deposited into pit latrines or cesspools located in backyards or at some distance from dwellings (because of the odor). The accumulated waste was then transported to nearby farms for use as fertilizer. Before the nineteenth century, a city's open "sewers" were used to disperse rainwater and urban detritus. The development of pipes bringing water into the residential buildings of nineteenth century Europe, North America and other wealthy countries also led to a huge increase in water consumption: from 5 gallons (20 liters) to 30-50 gallons

(120-200 liters), per person, per day. So, once indoor plumbing came along, the backyard privy or cesspool was now fed with shit washed out through pipes, using copious amounts of fresh water. This inevitably led to overflows, a horrible stench and thence, a desperate search for a solution.

The first fix was the use of open sewers to transport sewage away from population centers. This resulted in deadly outbreaks of water-borne diseases like cholera and typhoid. The next fix was to create a network of pipes to protect the population from the pathogens in raw sewage. But where should all the sewage go? Argument raged between those who wanted to send it to fields as fertilizer, and those who favored sending it to the closest river, lake or coastal water, for disposal. It was won by the latter. The mantra was "running water purifies itself." Not really, and not sufficiently for cities downstream which now had to use water that was seriously polluted. The next techno-fix was downstream cities filtering and then disinfecting the water with chlorine before using it. Now that previously unpolluted aquatic ecosystems were polluted with human waste and excessive nutrients, the need arose for yet another techno-fix: "treating" the wastewater. Thus evolved the modern sewage treatment plant, designed to clean and disinfect wastewater before discharging it into a river or ocean.

The Royal Loo – the velvet lined toilet of King Henry VIII at Hampton Court, near London.

Problems have arisen with this approach. The pollution of waters is now virtually everywhere. The reason is simple: Where do you send those millions of gallons of effluent—the liquid waste, increased in volume by the water needed to flush it away? The amount of water needed per human bowel movement can be as much as 5 gallons (20 liters) with old style toilets, down to 1 to 2 gallons (4-8 liters) with modern, water-conserving flush toilets. If 2 billion people are using a centralized sewage system and visit the loo twice a day, that's between 2 and 10 billion gallons (8 to 40 billion liters) of sewage water per day. If it takes one or two thousand tons of water to move each ton of shit, it's lucky that most cities are adjacent to a river, lake or an ocean.

In rural areas septic tank systems release wastewater which percolates down into the water table. Where the population is low, or where there are deep aquifers, there is little problem with this situation. Where the water table is closer to the surface or population density is high, groundwater is at risk of contamination. Few realize that even in developed, urbanized countries like the United States over a quarter of the population send their waste to septic tank and leach drain systems, not to centralized sewage treatment plants.

This huge waste of water is unsustainable. It would be one thing if the water used to flush was of low quality. It isn't. We use fresh, potable water. Human excrement contains 5-7% nitrogen and 3-5% phosphorus, two valuable nutrients, understood as critical for plant growth because of their relative scarcity and irreplaceability being flushed down the toilet in the name of hygiene.

I can't resist sharing an anecdote before we leave the topic of indoor plumbing. In the mid-nineteenth century, an Englishman named Thomas Crapper was marketing his new invention: an improved indoor toilet seat with a water siphoning system to regulate the amount of water needed for flushing. Previously, a vast amount of water was wasted since the amount used was determined by the user who turned on the water until it washed all the solids away. But now you could pull a chain and a regulated amount of water would flush your waste away. Thomas was rewarded, being appointed plumber to the Royal Family. It did not occur to him that his name and invention would be immortalized in the English speaking world. American soldiers, for example, who saw the name on English plumbing fixtures, during World War I, would say: "Excuse me, but I've got to go to the 'crapper.'"

Valveless Waste Preventer by Thomas Crapper & Co. of Chelsea, London, 1900.

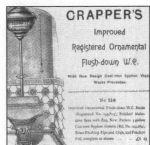

Advertisement for Thomas Crapper's improved flush toilet, late 19th century. Argument rages over whether he invented the siphon device that dramatically reduced the amount of water needed for flushing, but he certainly helped make it widely known and adopted.

The Solution to Pollution is Dilution

Why is dilution considered acceptable? Well, believe it or not, the adage of sanitary engineers of old was "the solution to pollution is dilution." In other words, they imagined that there's enough water in that river, lake, or ocean which will dilute the sewage we put into it, to the point where it's not a problem any longer. Consider the corollary in trying to solve air pollution problems. It would be to build smokestacks ever taller so that the pollutants in the air can't be smelled or deposited locally. Instead they would just get dispersed in the air currents of the upper atmosphere. The entire global atmosphere is thought of as large enough to make the problem disappear.

Another consequence of indoor plumbing and centralized sewage disposal was that it was decided that the same solution should be applied to industrial waste. This was very convenient for industry, since taxpayer dollars could be used to pay for removing its waste. So, in "advanced countries" pipes carry away both the relatively innocuous humanure and other wastewater from the family home as well as the chemicals from drycleaners, photo labs, food processing plants and factories. A dizzying variety of man-made chemicals have been thrust into our environment, including deadly pesticides as well as heavy metals. The presence of these compounds has made the disposal or reuse of sewage sludge (the solids removed at the sewage plant) much more difficult and hazardous as long-term health consequences are little understood. Application of sewage sludge on land as a fertilizer raises the danger of further contamination of soil and water, and of uptake by crops. Amazingly, this subsidy that industry enjoys and the environmental and economic costs of mixing human and industrial sewage, are rarely discussed. Separating industrial waste from domestic waste would now be very difficult since a huge infrastructure has been built in the world's cities, based on the notion that it was okay to mix them in the first place.

The thinking behind indoor plumbing and centralized sewage treatment plants was based on the ease with which one could dispose of the nasty stuff and a recognition of the health hazards associated with waste disposal. Contact with improperly treated shit can spread disease. These diseases are at devastating and increasing levels in the poorer, developing countries of the world. The "North," the developed world, suffers from problems caused by indoor plumbing, i.e., polluted water bodies and increasing difficulty in disposing of industrial sludge and the need for expensive sewage disposal infrastructure. The "South," the developing world (formerly called "the Third World") faces a different problem. The United Nations Development Program (UNDP) estimates that more than a billion people lack access to clean drinking water and three billion people lack access to adequate sanitation. In developing countries, 95% of the sewage is discharged untreated, polluting groundwater, rivers, lakes and coastal areas. Since groundwater, rivers and lakes supply drinking water, the consequences to health are dire. Over a billion people in these developing countries suffer from diseases caused by contaminated drinking water. The number of people who die each year from waterborne diseases totals three and a half million, mostly children under age five. That's more than 9,000 people a day who die from illnesses caused by the pollution of water. Diseases caused by sewage contamination of drinking water are by far the world's greatest killer.

Sanitary engineers of the nineteenth and twentieth centuries can rightly point to an improvement in human hygiene due to the use of indoor plumbing and sewage systems—a vital factor in better sanitary standards, a reduction in the spread of infectious disease, a decrease in child mortality and an increase in life expectancy. But it's now time for a paradigm shift in thinking and implementation of effective but more ecologically attuned solutions.

Human shit is not a toxic waste product and should not be treated as if it were. Throwing away vast amounts of potable water is increasingly untenable in a world facing a shortage of fresh water. Solutions to health and fresh water issues in the developing world do not have to replicate the energy consuming, wasteful technologies which were adopted by Western countries. The costs of building and maintaining centralized sewage treatment plants are not only enormous, the world's supply of freshwater will not permit extending these wasteful practices even to the world's *current* population. Freshwater shortages and increasing water pollution are making it clear that developed countries too cannot afford to continue such practices indefinitely.

Planet Earth is two-thirds ocean. As the great inventor and thinker R. Buckminster Fuller noted, it should really be called Planet Ocean. However, 97% of our planet's water is salty and not suitable for drinking or for irrigation. Most freshwater exists as ice, in the polar regions. Potable water is a precious and scarce resource.

But we are not "up the creek without a paddle." In this book, I will take you on my thirty year odyssey around the world, looking for ways out of this shitty mess.

View of Synergia Ranch, 20 miles southwest of Santa Fe, New Mexico as it looked in the 1970s. The photograph shows the geodesic dome, adobe buildings and animal yards, with the Ortiz Mountains to the south. The Ranch, situated in semi-arid high grasslands with piñon and juniper trees, had been desertified by overgrazing and soil erosion, which is why there are gaps of bare soil between the pasture grasses.

2
Flower Power at Synergia Ranch

"Money is like shit.
Does a world of good if you spread it around,
but pile it up in one place and it sure stinks!"
—Texas saying.

I N 1968 I GRADUATED summa cum laude from Dartmouth College. Resisting my Jewish Eastern European immigrant family's dictum that all males of my generation should become medical doctors, I decided to seek a different, more personally fulfilling destiny. It was the 1960s, and I was mindful of the slogan: "You're either part of the solution or you're part of the problem!" No disrespect to medical doctors, but I reasoned at the time that doctors basically patch up people suffering from unsatisfying lives to send them back into environmentally deadly and soul-stultifying jobs. I wanted to work on developing better ways of living, starting with my own.

The author pruning a newly planted fruit tree in the orchard at Synergia Ranch.

I had met someone who knew of an innovative group of people working on better ways, out in the wilds of New Mexico. She said they were "on the land" doing ecological work. I had a Volkswagen bug, a gift from my elder brother who had completed medical school and was doing his internship in New York, so I loaded my gear and headed southwest.

Synergia Ranch

Driving the last few miles to the group's "Synergia Ranch," twenty miles from Santa Fe, New Mexico, I had my first experience of driving on an unpaved road, the first and least of the culture shocks that were in store for me. As a boy in New York, I had often wondered if there was still soil under the paved surfaces.

The Synergia community was living on 165 acres of semiarid land, at an elevation of 6,200 feet. The site had once been reasonably productive grassland with a covering of juniper and piñon trees. First homesteaded in the 1920s, it had been overgrazed by sheep and was now in serious decline. The front and back pastures had been ploughed after the junipers and piñon trees were removed, and they were planted with wheat and corn. After a couple of crops, the bare and exposed land began to blow and wash away.

The Western United States was opened up by its first wave of settlers whose motto was "Rain follows the plough." In other words, hard work and enterprise would be rewarded with adequate rain. This had worked fine in the Midwest. Once its original forests were cleared, abundant rains and deep, rich soils were able to sustain good farm and pasture land. But, alas, in New Mexico and the Southwest United States, the soils were too fragile, the rains far less plentiful and erratic. So when this kind of land was stripped of its native vegetation, several centuries of land degradation and soil erosion ensued—a human caused desertification.

The land appeared ravaged. It was beautiful but it looked as though some terrible calamity had struck. The scattered grass clumps in the "pastures" stood up an inch or two above the soil. I

Looking south toward the homestead area of Synergia Ranch showing how the degraded land had been invaded by cacti, supporting little vegetation and mostly poor pasture species.

was dismayed to learn that the surrounding soil had been blown or washed away. The land was continuing to erode with every strong wind or downpour. There were signs of recent erosion: the *arroyos* (Spanish for dry washes or gullies) were getting deeper as the periodic torrential rains washed away the soil, since there was so little vegetation to hold it now.

John Allen, director of development, and Marie Harding, Synergia Ranch's manager and owner, were willing to let me stay and learn. Their minimum requirement was that I be "a willing incompetent." Willing I was and definitely! But I suddenly became aware how incompetent and impractical I was. I had never planted, grown, or built anything in my life. I was in a new world. People would say: "Hand me a 2 by 4" and I had no idea what they meant. Two by four what?

John Allen is a mining and metallurgical engineer with an MBA from Harvard Business School. He's a jovial man. When I asked what I could be responsible for during "ranch work" (which was part of the deal if you wanted to stay at Synergia Ranch), he said I could be in charge of "Gardens and Trees." There were vegetable

(Left) The old bunkhouse and kitchen at Synergia Ranch, early 1970s. The Ranch was stark and virtually tree-less when ecological restoration began.

(Right) Synergia Ranch today, looking toward old Ranch kitchen. Compared with the earlier photo, even the weeds have improved—now grasses volunteer under the shelterbelts of the trees, and contribute to holding and improving the soil.

gardens to manage and a plan was being developed for planting trees to hold on to the remaining soil, and also plans for an orchard of fruit trees.

I quickly learned how to determine where the roots end and the trunk or stem begins in a young tree seedling. As soon as I did, my success rate at planting windbreak trees soared. Now the trees had a fighting chance!

When I first started planting trees in New Mexico, we used a five gallon bucket of compost to give new tree seedlings a good start. By the time I left New Mexico in 1978 to launch a project in Australia, I had taken to digging a three foot diameter hole under every tree and filling it with maybe a half ton of compost. I had become a believer (at least in this case) that although small may be beautiful, the more compost you gave that tree, the better.

An Ode to Soil

The New Mexican "soil" was great for making adobe bricks. Just add a bit of sand and straw, mix with water, shovel it into wooden forms, let it bake in the sun and you've got adobe with which you can build buildings. The native tribes had been doing just that for several thousand years. But the topsoil was mostly gone, along with its nutrients like nitrogen and phosphorus. Further, the dry climate made the soil alkaline (pHs 8-9) thus chemically binding useful trace elements (iron, zinc) which otherwise would be used by plants. Dig down a foot or two and you hit caliche—an impenetrable layer of calcium carbonate. Nutrients were scarce and the sparse vegetation couldn't supply the soil with sufficient organic material. Yet soil itself is an "evolving being," as Russian soil scientists call it, unceasingly being made by tiny organisms and earthworms. Soil is a very complex material. One shovel full of fertile soil contains a greater number of living organisms than all the humans who have ever walked on Earth. One teaspoon of soil contains five billion bacteria, five million amoebas, thousands of fungi, tiny roots and hairs and other life forms most people have never heard of. Unlike what is washed out of clothes being laundered, soil is not "dirt." Soils are full of life. Life has transformed planet Earth; life is insatiable and unstoppable if you let it be. Turn over a rock in the most godforsaken spot, or check out underwater thermal vents or look under the ice in the arid valleys of Antarctica, and life will be found.

An Ode to Our Animals

Fortunately, we had potential fertilizer factories at the ranch: chicken coops, pig pens, horse corrals, and outhouses. Our animals produced plenty of fertilizer. Chicken shit is literally hot stuff. It is so high in nitrogen that it can "burn" plants if used directly on a garden. It was a precious additive for the compost heaps. From time to time, we kept other animals: some cows, goats, a donkey, sheep, and even llamas for wool, but horses—the largest of the domesticated animals we kept—supplied the most. In the course of my education in the wonders of human and animal excrement, I came across an astonishing fact. A horse produces around 17,000 lbs. (7500 kgs.) of manure, a year. That's almost 50 lbs. (22 kgs.) a day!

A milk cow rests in front of a new adobe building at Synergia Ranch. The adobe bricks were made on-site and the doors and windows were handcrafted in a woodshop at the ranch.

An Ode to the Outhouse

No one knows who invented the outhouse. Dig a hole, put a small structure with a toilet seat on top, add a roof and a door for privacy, and a roll of toilet paper. When it gets full, dig up the shit and compost it! Or you can simply plant a tree in the hole and move the outhouse to another location.

Outhouses are not perfect everywhere, but in New Mexico, they were nearly ideal. It is said that the danger of pollution from an outhouse (sometimes called a pit latrine) is about three feet in every

Aerial photo of the young orchard at the end of the 1970s at Synergia Ranch. Right of the orchard is the front pasture, denuded of juniper trees by dynamite (!) in the 1930s. To the left, is the outline of the micro catchment experiments and the juniper-piñon grasslands of back acres of the Ranch land.

direction. That might present problems if groundwater is close to the surface. But our groundwater was hundreds of feet down, so it was unlikely to be contaminated by anything leaching from an outhouse. It doesn't rain often at Synergia Ranch; and since there is no flushing system in an outhouse, no water is added so the nutrients in the humanure stay put.

Did I say outhouses were "nearly" ideal? There were freezing winter nights when you had to go to the unheated outhouse nearest your room. And it was fragrant. Though we used lime to take the odor away, we didn't have ventilating fans that modern composting toilets have, nor did we know enough to install a vent pipe to disperse the odors above the outhouse roof. It wasn't part of the Ranch lifestyle then to use air freshener to try to mask the obvious. It was a good basic training to bring over-educated urbanites down to earth. When you took the outhouse away and dealt with the hole, there was a strong smell. It was one of my responsibilities as head of Gardens and Trees to shovel out the outhouses when they filled up and haul the steaming contents to our compost piles. It's surprising what you can get used to–the 100% natural odor!

What startled me when I first started shoveling shit were the black widow spiders. I found them in great numbers under the toilet seats. There they were, just inches from where my friends and I would sit. Yet there was never an instance of anyone being bitten. Evidently black widow spiders have their own food sources and weren't interested in biting our genitals. Moreover, as the saying goes: "The more spiders, the fewer flies!"

An Ode to Horse Manure and a New Orchard

The Ranch was changing from a windblown but scenic horror. The trees we had planted gave each other protection. As they passed the survival stage and began growing, the change was impressive. Now, around a thousand young trees at Synergia Ranch were shading the ground and protecting the soil. Their roots aerated the soil. The trees created a microclimate around the ranch, adding moisture to the air as they transpired water. You could feel the difference after hiking in the nearby beautiful but stark Cerrillos Hills. Returning to the Ranch was like entering a moist, cool greenhouse. The soil, enriched by compost and protected by the trees, now supported ever more earthworms. They upgraded the soil. And we planted "companion plants" around the trees— alfalfa and clover to add nitrogen to the soil; herbs to repel insects. Even the quality of our weeds improved. Native flora like the gramma grasses started superseding some of the lesser value vegetation that were once the only plants that could survive in the poor, eroded soil.

Marie Harding and Chili Hawes apply composted manure in the orchard.

I visualized an orchard with 425 trees, set on the east front pasture of the ranch. The soil there was good, with two and a half to three feet of clay above the caliche. The land sloped away from prevailing winds and offered some protection from spring frosts.

For my dream of a fruit tree orchard, the outhouses and our animals would not be able to supply enough manure or compost. Fortuitously, a horse racetrack had opened just eight miles from the ranch. I could rent a dump truck and take advantage of the twelve hundred racehorses there. The track was creating a small mountain of horse manure where they were dumping their precious waste. The track's owners were more than happy to supply me with horseshit. I put on a cowboy hat and went and closed a deal.

To put in our fruit trees, I engaged a friend who owned a backhoe. The assignment was to dig holes three feet across and three feet down so I could place compost below the roots of the newly planted trees. They'd have a decade worth of the nutrients and water-retaining soil they needed.

I'm not sure who first gave me the new nickname. I became known as "Horseshit." I reckon I hauled 500 to 600 tons of horse manure from the racetrack, back in 1974-75, both for the new orchard and to feed all our other homestead trees.

I was, at least for that season, the horseshit king of New Mexico. I was also managing a landscaping company at this time, having sold my interest in the Ranch woodshop, and we were landscaping a number of houses in Santa Fe. People were always happy to see me. I can still hear their welcoming cries of "Hey, Horseshit!"

Ode to the Earthworm

Charles Darwin, best remembered for his work on the theory of evolution, was also a great fan of earthworms. He went so far as to assert that earthworms were perhaps the most useful species on the planet. He observed what they were doing for the soils of Great Britain. He calculated that, given their numbers (millions per acre), and their long workdays of digging through and consuming earth and then producing organically-enriched compounds, earthworms were making one inch (two and a half centimeters) of topsoil per century. That adds up.

Of course, to support large populations of earthworms and to benefit from their industry, conditions have to be right. Earthworms can only thrive if there is enough moisture and adequate nutrients in the soil and plant cover on the surface.

New Mexico was bereft of good soil but at Synergia Ranch we asked ourselves, if earthworms can make topsoil, why can't we Homo sapiens?

An Ode to Compost

People have been making soil for centuries. The process is called composting. After seeing it at work in India in the 1940s, Sir Albert Howard brought the technique back to England and started the modern "organic farming" movement. His insight was that healthy soil is vital for crops and ultimately human health. Since then, composting has evolved in many directions and there are numerous books on the subject, taking different approaches. But, no matter which approach is used, composting remains the key way that human and animal manure can be rendered safe and then transformed into topsoil.

In our early years in New Mexico, we followed some of the guidelines of the biodynamic gardening approach as well as the more popular Rodale method of organic farming. The basic formula for compost was to mix together roughly 70% crop residues / straw / kitchen scraps with 20% manure (of whatever origin) with 10% soil. Manure provides the nitrogen which heats up the mix. The soil supplies the microbes and other living organisms which make more live soil, just like a "yogurt starter." The vegetable

matter provides bulk and helps keep the pile from getting compacted and losing its aeration. It's important that the manure be as fresh
as possible, because old stuff loses a good percentage of its nitrogen.
A good compost pile can heat up enough, over a few days, to burn
your hand. Temperatures of 120-160 degrees Fahrenheit are not un-
common. That heat not only kick-starts the transformation process,
it will also kill all the pathogens, parasites and viruses and the seeds
of weeds. You don't want the temperature to get too high; otherwise
you can lose nitrogen and even destroy microbes essential to the
later stages of the compost process. Making sure the pile is sufficiently moist is the key to preventing overheating. There have been
incidents where municipal compost heaps spontaneously combust!
A compost heap should be as moist as a saturated sponge but not
so wet that air-breathing microbes are drowned, which would make
the compost anaerobic, causing a stinky odor and slowing down the
process. After the initial heating, a compost pile cools down and the
time is right to add earthworms to assist the transformation of the
waste into new, organic-rich soil.

Another good idea is to have neighboring compost piles started
at different times. The worms know when the temperatures drop
and it's okay for them to move into neighboring compost piles.
And, if you keep adding to a compost heap, there will be hotter,
fresher sections, and worms will move to the older sections, moving
back when the heating phase is over. Compost worms are a red
hybrid variety of earthworms that thrive in compost heaps. They
multiply at a faster rate than ordinary earthworms. They are grown
commercially for fishermen so they are not hard to find. In New
Mexico, because of the dry climate, we usually covered our compost
heaps with a layer of soil or an old carpet, and our main composting
area was dug a couple of feet into the ground to prevent it from
drying out. Whenever you add anything smelly, like human waste

or food scraps, make sure the compost heap is covered with straw or soil. This will prevent odors from escaping and deter wildlife. A well-covered compost heap keeps flies away, too.

So what takes earthworms a century to create can be achieved in a matter of months using stuff that is normally considered useless. This isn't simply recycling—this is transformation. That unpromising pile of waste material has turned into sweet-smelling, black topsoil.

There are composting methods which call for turning over the piles in order to speed up the process. That seemed like a lot of effort, especially at Synergia Ranch as we didn't then have a tractor. We did all our gardening by hand. Research indicates that the more you turn a pile the more nutrients are lost. The aeration is short-lived. Aeration is best achieved by including lots of high carbon vegetal material, like straw, hay, sawdust, or crop residues (the stems and inedible parts of crops). So, we were actually smart to adopt a more relaxed routine in our compost making. We started a lot of compost piles in the spring so that they would be ready for spreading on the garden in the fall. Similarly, compost started in the fall would be ready for adding to the soil at planting time, the following spring.

"What? You used human shit on your garden?" Pathogens are killed during the heating of a compost heap. But, as an additional safety measure, we made separate compost heaps using humanure or animal manure. The "human compost" was earmarked for use when planting trees, for fertilizing under their leaf canopy; so there was no danger of contaminating what was grown in the vegetable garden. Our vegetable gardens received animal manure based compost. This was mixed with coffee grounds and kitchen waste and other vegetable waste. Also, I had started a woodshop and the sawdust from furniture we made was put to good use. Pine sawdust is acidic so it helped make the compost even more effective at neutralizing, thus correcting the alkalinity of our soil.

An Ode to the Trees

We planted over a thousand shade and fruit trees around the central homestead, including drought-resistant trees like Russian olive, green ash, lilac, honey locust, Chinese elm, native plum along with apricot, cherry, peach, plum, apple and pear fruit trees.

Trees for windbreaks were supplied to us by the New Mexico State Forestry Service at the laughable cost of $5.00 per 100 seedlings. They were planted with five gallons of manure or compost and then zealously watered. With much hard work they survived and even grew a little the first year or two. Eventually they adapted to the harsh, semi-desert climate.

Our water was pumped from 280 feet (86 meters) below the ground and it was loaded with minerals. When the water evaporates it leaves a deposit of white powder—calcium and magnesium carbonate. That was the stuff which made the soil so alkaline. Our soil didn't need any more salts! So, we soon became diligent about watering at night, using drip irrigation to minimize the amount of water lost to evaporation. A protective mulch layer of pumice, straw and compost helped prevent a buildup of salts and helped improve the moisture-holding capacity of the soil underneath.

Compost is stabilized black soil. It is more than a temporary fertilizer and sun block. Manure applied to soil provides nutrients for a season but eventually they leach away. Compost is more than twice as effective as manure and it has the same moisture-retaining capacity as topsoil. It is estimated that compost can hold more than five times its weight in water. That was additional protection for our trees. It meant they wouldn't dry out so quickly and we could reduce the amount of irrigation they needed.

John Allen understood that a common mistake people make in dry, semi-desert areas is to give their trees frequent but shallow waterings. This leads to evaporation and salt buildup. Shallow waterings also condition roots to be lazy: they don't search for more water, deep down. So we stretched out the intervals between waterings and the trees got taller. We watered longer but less often, to force the roots to go deeper. In order not to harm the gardens or the trees you had to pay attention and distinguish between a temporary "day wilt" and the point where plants could be permanently damaged by lack of water. It worked! Our trees became robust and developed deep root systems. And the more compost we added to the soil, the more moisture the soil could absorb and make available to the trees in between waterings.

An Ode to Water and Desert Wisdom

During my New Mexico apprenticeship in desert farming, I was able to learn from two gifted desert farmers. One was a Hopi Indian, the other an Israeli scientist who was employing the ancient methods of the Nabateans, in the Negev Desert.

Leslie was an Elder at Sipaulovi, one of the Hopi villages in northern Arizona. I had an introduction from the French mystic Robert Boissier, who had married a Hopi woman and moved to Santa Fe. Hopis call white men "Bohanna," and regard most white men as insensitive jackasses. Seeing what first the Spanish and then the Anglos did to the richness of the Southwest, who can blame them? Robert Boissier's running joke was that he was from the "Banana Clan" (Hopi villages have two clans, each with their own sacred place, called a kiva). He instructed me to take some bananas to Leslie as a kind of coded signal, when I made contact.

I found Leslie in a simple lean-to structure. It was midday and I announced that I was willing to work and do whatever he told me. He looked at me with a mixture of amusement and sympathy: "Rest first, sun high. When sun low, we work."

Water catchment in early planting of orchard.

He was a traditional farmer—raising corn, squash, beans and apricots, which were once the mainstay of the Hopi way of life. Now he farmed to have enough for use in ceremonial dances. In a land that only gets four inches of rain a year, Leslie grew crops that could cope with the harsh conditions. He planted his multicolored corn when winter snow was still on the ground, putting the seeds deep. By the time they sprouted, they had extensive root systems. He also spaced his corn several feet apart, and weeded out competing vegetation so that the corn would get all the rain.

One day, I found Leslie carefully stirring a mixture, in a small tin can. I watched as he daubed a little of it on each squash and bean plant in his desert garden. "It's to keep the rabbits away," he explained. "Special mixture?" I asked. Leslie laughed. "It's dog shit. Mix it up with some water. Rabbits never come near the plants when you put this on. Gotta put it on again though if it rains." I wonder why Dow Chemical and Monsanto haven't patented this?

I went to the Negev Desert, in Israel, after reading a book that the Israeli scientist Michael Evenari and his team wrote about techniques in rain harvesting. Here, in a really inhospitable climate, cities had flourished and farms had supplied them as well as traders traveling the caravan routes. Evenari, a biologist at Hebrew National University in Jerusalem, had shown that the ancient Nabateans had diligently collected every drop of rain that fell on hillsides and had sculpted channels which took the water to the valleys where they had laid out their fields. The same farms, reinstated in the 1960s, worked perfectly—growing fruit trees and crops.

When it does rain in desert or arid regions, it is often a quick downpour and poor soils cannot absorb the rain quickly enough so there is runoff. So, instead of letting the water go everywhere, the farmers of old constructed watercourses lined with stone to guide the rainwater down to the valleys. There, the loamy soil can retain

The author in front of peach trees in the Synergia Ranch orchard, late 1970s.

the water. The Negev Desert gets about 4 or 5 inches of rain a year, mostly in the wintertime. The soil retains enough moisture to grow crops in spring and summer. Twenty to thirty acres (8 to 12 hectares) of hillside channel enough rainwater to provide for one acre of valley farmland.

I took these methods back to Synergia Ranch and started experimenting. The new orchard thrived and the windbreaks grew and flourished. Life begets life.

These experiences were my initiation into the transformative power of so-called waste, which rightly applied to the soil, helps to sustain life. In 1973 my friends and I decided to start an organization for the study of ecological processes. We called it the Institute of Ecotechnics. We believed that a harmonization of ecology and technology was the great challenge of our time. We also determined that we needed some prototype demonstration projects to test our theories. We decided to go global and start projects in different

A 50 foot (15m) diameter, 27 foot (8m) high geodesic dome with canvas covering was built at Synergia Ranch in 1971. This photograph shows common lilacs, Persian lilacs and elm trees which were planted as part of the greening of the ranch; ground cover is increasing and native grasses are starting to flourish.

regions around the world, from the Mediterranean to the Tropics, in the rainforest and even at sea, onboard ship. We were especially interested in the health and welfare of coral reefs. We recognized that people and their cultures, from their art to their waste products, are a powerful yet not properly integrated element of the planetary scheme of things.

It seemed crucial to work out how communities living on the land can increase biomass and diversity while also satisfying their economic needs. At Synergia Ranch we learned how to make an oasis from a previously desertified environment. An intelligent integration of the people and their technologies led to an increase in the land's potential. Henry Louis Le Chatelier, a French scientist of the eighteenth century, formulated a most important principle. Any system, including a living ecosystem, has a direction. Systems in decline continue to decline and systems that are improving, continue to improve—unless enough countervailing forces are applied. The early years in New Mexico were tough. The system was in decline. It only took one mistake to lose a tree. Later, the system reversed its direction and the default direction shifted.

Apple blossoms on one of 440 fruit trees in the orchard.

At first you work and work hard, yet little seems to improve. Yet over time things do change and improve. There were many factors that contributed to the transformation at Synergia Ranch, but the primary one was harnessing the potentiality of shit to make rich soil.

Historically, humans have often been unintentional desert makers. This led to the demise of many civilizations. There are haunting images of desert sands covering the ruins of once great cities. We are not doomed to repeat the mistakes of history. James Joyce's memorable line in *Portrait of the Artist as a Young Man* is apt: "History is a nightmare from which I am trying to awake." We can learn to make oases, not deserts, and intelligent recycling of human shit is a key part of the solution.

Biosphere 2 was the largest laboratory for global ecology ever built. A few months before September 1991 (the start of the first two-year closure experiment), I was one of eight people selected to spend two years living inside Biosphere 2.

3

Recycling in a Small World: Biosphere 2

"There's just one place that a man can be alone, even on his wedding night."
—from a song in Bertolt Brecht's play *Baal*, referring to the solitary joys of the outhouse.

From the Outback to Space Biospheres

In 1985 I moved to Arizona, after seven years based in Australia where I had helped start a tropical savannah restoration project. My colleagues and I were now undertaking a wildly ambitious and visionary project: an experiment to create the largest scale closed ecological system with humans ever attempted. It was to become the first man-made biospheric system—Biosphere 2.

Biosphere 2 was the summation of all our previous work in ecotechnics. Our institute was registered in England, with offices in London. This new project of creating a completely self-contained world was going to need inputs from engineers, ecologists, agriculturists, entomologists, botanists, architects and builders, attracted to the challenge of doing something that had never been done before. Biosphere 2 would put it all to the test.

This is not the moment to discuss in detail the technical challenge of making the massive structure of Biosphere 2 airtight or the feat of collecting plants, animals and corals from around the world, the fabrication of appropriate soils or designing technologies to create diverse and functioning ecosystems. Biosphere 2 included areas that replicated

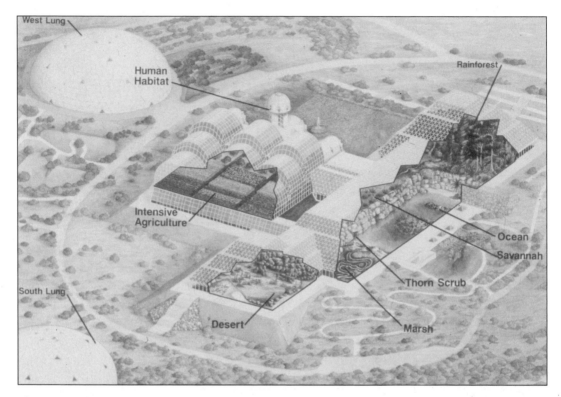

West Lung

Human
Habitat

Rainforest

Intensive
Agriculture

Ocean

Savannah

Thorn Scrub

South Lung

Desert

Marsh

Schematic showing the major components of Biosphere 2: the wilderness biomes including rainforest, savannah, thorn scrub, desert, marsh and coral reef ocean, the two anthropogenic biomes: agriculture and living space (human habitat) and one of the two "lungs" (variable volume chambers). Biosphere 2, the most tightly sealed building in the world, was ecologically "another world," where recycling and making renewable use of all resources was a necessity, not a choice.

natural biomes: rainforest, savannah, thorn scrub, desert, fresh and saltwater marshes and an ocean with a coral reef. Another section was a mini-farm for intensive agriculture and the human habitat (a kitchen, labs, workshops, recreation areas, crew's private quarters, and offices). Air, water and nutrient recycling systems were completely interconnected. From the perspective of human waste what's essential about a materially-closed ecological system is that what's inside remains inside and is forever recycled.

I had made contact with some leading Russian scientists who were experts in the field of closed ecological systems when I'd stopped off in Moscow en route to Australia. When I got Down Under I sent them all boomerangs as gifts. Just like a boomerang, everything in a closed system keeps coming back. And the cycle is

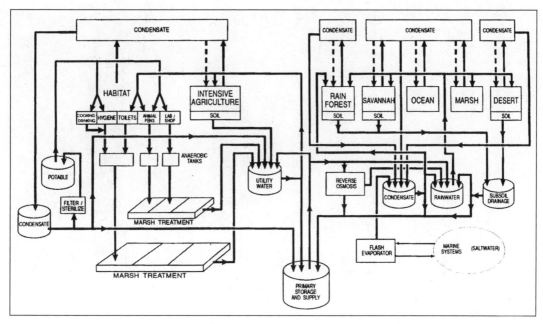

much faster because no matter how large a man-made biosphere and life support system is, the buffers of water, air and soil are much smaller in a man-made system. The concentration of life is much greater. We had calculated that in Biosphere 2 a carbon atom would recycle, on average, in about four days. That is, from being part of a molecule of carbon dioxide in the air, then being used by green plants in photosynthesis, respired by microbes, plants or animals or absorbed and then released from the mini-ocean to take up residence again in the atmosphere. Such a cycle takes three to four years on Earth. In Biosphere 2 it would be a thousand times quicker. The same was true of the water cycle. Water inside the facility would be recycled and become drinking water again, within weeks. "Instant karma!"

In such a sealed system, it becomes untenable to use toxic chemicals. All infrastructure (structural components, machines, materials) had to be checked to see if any of these released unwanted

Schematic of the water systems inside Biosphere 2, including the wastewater treatment system. The constructed wetland helped complete the water and nutrient recycling; treated water returned to the irrigation supply for the farm, and nutrients contained in the vegetation were cut and fed to the domestic animals or composted, thus returning nutrients to the soils. Drinking water was condensed from atmospheric humidity and was very pure because of all the green plant transpiration.

The Test Module, center of our closed ecological system research (1986-1989), while Biosphere 2 was under construction. My twenty-four hours in the Test Module helped me decide to become a biospherian and begin the training necessary to run a small world.

Interior of Biosphere 2 Test Module. On the lower right is the small constructed wetland for recycling wastewater from one inhabitant.

compounds, or ones that couldn't be broken down and reused, into the cycle of this mini-biosphere. Biosphere 2 set a world record in terms of almost no leakage—less than one percent per month, despite its 3.2 acre (1.2 hectare) size, ceilings up to 75 feet (23 m) and miles of space frame.

You could never think of throwing something away. Firstly, there was no "away" since you could walk from one end of the small world to the other in fifteen minutes. Secondly, if that something was made of elements vital to life (carbon, oxygen, nitrogen), it was necessary to keep it circulating; otherwise, the system could accumulate what the Russians called "dead-lock substances" (substances that can't be recycled). Even in the most advanced Russian closed systems, a portion of the necessary protein, i.e., meat was imported and human waste was exported. In Biosphere 2, we aimed to grow a complete diet and recycle all our wastes.

How would we handle our shit? An outhouse was out of the question! Besides, we had graywater, animal manure and urine, laboratory wastewater, inedible crop waste and a host of other stuff to

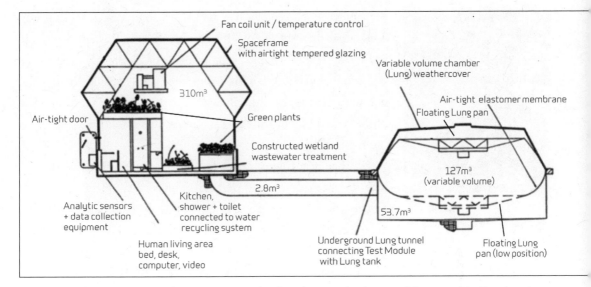

Fan coil unit / temperature control

Spaceframe
with airtight tempered glazing

Variable volume chamber
(Lung) weathercover

310m³

Air-tight elastomer membrane
Floating Lung pan

Air-tight door

Green plants

Constructed wetland
wastewater treatment

127m³
(variable volume)

2.8m³

Kitchen,
shower + toilet
connected to water
recycling system

53.7m³

Analytic sensors
+ data collection
equipment

Human living area
bed, desk,
computer, video

Underground Lung tunnel
connecting Test Module
with Lung tank

Floating Lung
pan (low position)

recycle. To find answers to this question and a few thousand others we decided to first build a test module and consult with experts around the world.

Schematic of the Biosphere 2 Test Module.

Biosphere 2 was intended to be both a laboratory to study global ecology as well as a prototype for large-scale biospheric life support systems that would be needed if and when humans start to live on other planets. We were surprised by the intense interest in our project from media around the world. Maybe people saw in Biosphere 2 an optimistic vision of how humans can live on Earth sustainably, and perhaps one day take life to other planets.

The Biosphere 2 Test Module had as much room as a medium sized apartment, one with high glass ceilings! It was made of steel panels laid out on a Buckminster Fuller-type space frame geometry. It had dozens of electronic sensors that kept tabs on key equipment, took periodic water and air samples and monitored anything else considered crucial to the health of the person and plants inside. Although the facility had less than one percent of the volume of the yet to be completed Biosphere 2, in 1987, the Test Module was

Dr. Billy C. Wolverton, pioneer in the use of constructed wetlands for wastewater treatment and in using plants for prevention of indoor air pollution.

the largest closed ecological system that had ever been built. It was a bit larger than the famous Bios 3 unit that the Russians had built at Krasnoyarsk in Siberia, at the Institute of Biophysics, where they had conducted research with two to three people inside, for months at a time. At the time, the Russians' work was years ahead of what NASA was doing on closed ecological systems and bioregenerative life support.

We had heard about a maverick NASA scientist, Billy Wolverton, who had been developing wetland systems to clean up toxic waste and recycle sewage. We found him in his laboratory at the NASA Stennis Space Center, in Mississippi. Wolverton agreed to collaborate on the recycling systems we would try out in the Test Module.

Billy had pioneered using plants to clean indoor air and had built wetland systems to handle water pollution of all kinds. He was a quiet and determined revolutionary and he liked to talk about his work in terms of a real green revolution which used plants to solve environmental issues.

When I went to see him at Stennis, I found him about as far away from the center of operations as possible. Billy didn't mind. Stennis is the NASA center where spacecraft engines are tested. Billy loves to tell how he convinced the director at Stennis to try his green plant system to clean water polluted with heavy metals that resulted from testing rocket engines. "Nobody believed my system would work. It was too simple—just wetland plants and microbes, like you find in swamps. When the director asked chemical engineers for a solution, they proposed a system that would cost millions of dollars. My system was cheap in comparison. So the director said, 'Let's give it a shot.'" Sure enough, Wolverton's wetland system proved remarkably efficient at removing heavy metals from waste-water. The constructed wetland prevented pollution that might also have threatened other water resources in the Mississippi Delta.

Constructed wetlands are good at dealing with all kinds of pollutants, from mine tailings to food waste. Such systems sometimes do have to include some preliminary detoxification of the pollutant to a point where the wetland can be effective. By contrast, it is much easier for wetlands to cope with the relatively benign organic compounds found in human waste. Wolverton's work was part of a paradigm shift in thinking. Wetlands, which have so often been drained, were being recognized as immensely valuable ecosystems and were gaining protection. Once regarded as worthless swamps because they can't support agriculture, wetlands are now called the kidneys of the planet because they are so efficient at removing harmful compounds before these reach rivers, lakes or the ocean. Natural wetlands support unique vegetation, provide habitat for many animals and birds, and are home for many aquatic species.

Billy Wolverton consulted on the design of our first mini-wetland used to treat wastewater in the Test Module. Since it was designed to support just one person, the constructed wetland was tiny. It measured less than twenty square feet (2 square meters), and had two compartments. All kitchen, shower, sink and toilet waste from the compact living area was first held in a small tank (which functioned as a septic tank) and was then fed into the wetland. There were no pumps, just gravity to move the wastewater.

The wetland system had mini-channels of water and an area of soil. This sewage treatment system contained a diversity of plants—floating plants like water hyacinth, which makes a beautiful purple flower; and plants with their roots in the wetland soils, like canna lilies, with their gorgeous orange and yellow flowers and bulrushes and cattails.

The system worked extremely well. Plants grew rapidly. There was one problem, however. We were only scheduling experiments with a human occupant occasionally, so the wetland looked forlorn and hungry when no one was using the toilet! It became standard procedure for someone to go into the Test Module when they had a call of nature, in order to keep our wetland plants happy.

I was Director of Environmental and Space Applications for the parent company building Biosphere 2, Space Biospheres Ventures. Biosphere 2 was planned as a one-hundred year experiment and we had a group of people in training for the first closure experiment. I thought I might also volunteer.

In 1989 I took part in our program that gave a number of key managerial staff the opportunity to live in the Test Module for twenty-four hours. We'd done a number of experiments including humans who half-jokingly became known as Vertebrate X, Y, or Z. Vertebrate X was John Allen, the inventor of Biosphere 2. He had spent a day inside and had waxed poetically about feeling at one with the living system. In later experiments, two future Biosphere 2 candidates, Gaie Alling and Linda Leigh, spent five days and twenty-one days inside, respectively, getting all their food from the system and all their wastewater recycled through the constructed wetland. After intense concerns about what might go wrong, we were now getting reports of how satisfying and uplifting an experience it was to live in this alternate world.

To live inside the Test Module, even for a day, was amazing. From the moment the metal airlock swung shut behind you and you took your first breaths of moist, tropical, sweet air, you became an inhabitant of a new world. One organism, one species amongst many: An animal metabolically feeding carbon dioxide to the plants, which in turn are supplying you with oxygen. It's something to experience—the green plants are your third lung. You look around your little world and thank them all.

The Test Module had a small kitchen garden which supplied grain, vegetables and salad greens. There was a tiny rice-growing

The "biospherian handshake." The author inside Biosphere 2, being visited by two esteemed friends–John Allen, inventor and Director of Research and Development at Biosphere 2, and Academician Oleg Gazenko, Director of the Institute of Biomedical Problems, Moscow, one of the greatest space physiologists.

pond from which an occasional tilapia fish could be caught, but the diet was mostly vegetarian. No matter, the food was fresh and dinner for one could easily be prepared in the tiny kitchen. Space was obviously very limited. There was a foldout Murphy bed attached to a wall. Friends outside said that as soon as I was inside the Test Module, I appeared relaxed and content. After I'd been to the bathroom, I would sometimes go over to the wetland to commune with plants that were now ingesting my modest contribution to the nutrient cycle. After the wastewater was purified in the wetland, nutrients and water would go to the irrigation tanks that fed other plants in the Test Module. This was my introduction to living with wetlands and it made it ever clearer to me that life thrives on so-called waste. It's a valuable raw material, not something to be regarded as waste to be thrown away.

One of my specific duties was to manage the sewage system in Biosphere 2. This was a transparent system. You could literally follow the pipes and see where the wastewater came from and where it went.

Biosphere 2 had two treatment systems. One was designed to handle laboratory wastewater and the urine and other soluble wastes from the animal pens. The other wetland system treated the wastewater from the toilets, showers, sinks, kitchen and laundry, in the human habitat. Both systems had pipes leading to the basement and a series of holding tanks, each with a capacity of around 250 gallons (1000 liters). These were the equivalent of septic tanks, where solid waste could be separated and digested by anaerobic bacteria.

Wastewater was then drained into two rows of three interconnected fiberglass tanks, which housed the constructed wetlands. They were deliberately situated next to space frame windows to give the plants the sunlight they needed. The wetlands were designed a bit like a meandering delta, with open waterways planted with floating plants, and riverbanks of soil, where other plant species were placed. Water was pumped back to the first fiberglass tank after it reached the third tank. When we needed to make room for more wastewater, some of the treated wastewater was pumped into a final holding tank while one of the initial holding tanks was opened to drain wastewater into the constructed wetland. Wastewater pumped out of the wetland would be mixed with other water. Water for irrigating the Biosphere 2 farm was a mix of treated wastewater from the constructed wetlands, leachate (water collected from what drained through the soils) and condensate (very pure water collected from the humidity of Biosphere 2's atmosphere).

The farm provided fodder for the animals and food for the biospherians. Inedible parts of crops were composted with our domestic animal (goats, chickens, pigs) waste. Wetland plants were also harvested for animal fodder or were composted. So, by means of compost and wastewater, all nutrients from the farm returned to their soils to begin the cycle again. Sustainability, anyone?

Biosphere 2's wetland system covered an area of 450 square feet (41 square meters). We used to joke that "the unmeasured life is not worth examining," updating Socrates's saying that "the unexamined

Biosphere 2 constructed wetland system. The wetlands were housed in a linked series of fiberglass tanks, supporting floating and rooted wetland species. This photo was taken shortly after planting; later the vegetation became taller and denser.

life is not worth living." This was in recognition of the fact that when you build an artificial world you have to think of everything, and it seems as though you are measuring and monitoring everything. Since Biosphere 2 was first and foremost a scientific experiment, quantifying and tracing the interconnections were crucial to understanding its dynamics and complexity.

With a fair degree of certainty I can state (having published scientific papers with these data) that we processed 250-290 gallons (1000-1100 liters) of wastewater a day through our wetlands. Most of it came from us human inhabitants, who used about thirty gallons of water, per person, per day. That's not so bad since an average American household uses up to 100 gallons (400 liters) per person, per day. Around 25 gallons (100 liters) of water came from animals and our laboratories sent us 10 gallons (40 liters) a day. Constructed wetlands take longer to process and recycle nutrients than conventional sewage treatment plants. In Biosphere 2, in order to treat human sewage effectively, wastewater needed to stay in the system for about four days.

A field of wheat nears maturity in the agricultural section of Biosphere 2. The farm was a high-yield, non-chemical and highly diverse experiment in sustainable farming. Nutrients from treated wastewater coming from the constructed wetlands were added to the irrigation supply and returned to maintain soil fertility.

But these numbers don't tell the personal story of what it was like to live with and manage a low-tech, "green" sewage system. How enjoyable it was to understand where the sewage was going and how much vital recycling the wetland was doing to maintain the health of my world.

Biosphere 2's wetlands were far from a monoculture. As small as it was, we had around fourteen different types of plants in our constructed wetland. Wetlands are a special form of habitat and they attract animals that enjoy a walk on the wet side. Our wetland supported geckos, green anole lizards and even a Colorado River toad. The toad was a stowaway for it was not deliberately introduced into our ecosystem by the designers. It somehow found its way inside when Biosphere 2 was being built. The wetland was also a favorite breeding place for our ladybugs. Ladybugs are considered "beneficial insects" because they eat insects that attack crops, so they were included as part of our non-toxic "Integrated Pest Management" approach to intensive agriculture.

Becoming subsistence farmers was one of the great challenges for me and my fellow urban, middle-class biospherians. Despite living on what we grew in Biosphere 2, for a week at a time, during pre-closure experiments, the reality of "you don't grow it, you don't eat it" took a while to hit home. Furthermore, our food production was initially affected by unusually cloudy weather caused by two years of consecutive El Niños. I was always on the lookout for fodder for our hungry chickens, pigs and goats. What a pleasure when I realized that greenery being produced in the wetlands could serve a dual purpose. I could harvest azolla, a water fern with high levels of protein, and water hyacinth and periodically cut canna and the bulrushes and cattails. The more fodder, the more eggs, milk and meat. What the animals wouldn't eat (and goats can be picky) went to the compost heaps. The compost heaps received all the inedible

The first step of the constructed wetlands: holding tanks which functioned like septic tanks for separating sewage solids from liquid. In these sedimentation tanks, anaerobic bacteria begin the breakdown and transformation of sewage. From these tanks, the liquid was pumped to the constructed wetland tanks filled with plants.

crop residues and the solid wastes from the animals, in turn making rich topsoil for the farm. It all helped put food on our dining table.

We did have an ultraviolet lighting apparatus which could be used to disinfect wastewater from the wetland before it went into irrigation tanks. Apart from checking from time to time that it worked, the device was not used during my two years inside Biosphere 2. Pathogens are only a concern in sewage if the producers have any diseases. We eight biospherians had been medically worked over almost as much as astronauts. Not only didn't we have any diseases; there was no way any could get into our world.

That did allow me to be a trifle cavalier about physical contact with wastewater in Biosphere 2. In order to prune the farthest plants I had to balance on the edge of the wetland tanks. Occasionally I had to put on my swimming trunks and take the plunge into the partially purified wastewater.

I wasn't worried about germs but I was a bit timid about being seen cavorting half naked, in front of the thousands of visitors who came to Biosphere 2. The constructed wetland was completely visible through the well-sealed glass. Tour groups came right up to the windows. I knew that there was a sign outside which explained the purpose of the wetland but during the first few months inside, I did my sewage treatment work early in the morning, before any tour group came by.

Later, I began to enjoy seeing people share in the fun of living in Biosphere 2. I noted the incredulity evident on the faces of visitors, as the tour guide explained that this green oasis was also the sewage system. No one was in the habit of paying good money to visit a sewage treatment plant, let alone one with beautiful plants and sometimes a determined Jewish boy, sickle in hand, wading through the wetland, merrily filling his buckets with vital greenery.

As used to this routine as I became, I knew there had to be better ways of designing wetlands. This one worked, but was far from being perfectly problem-free, not to mention the odor. My colleagues disagreed over this issue. Probably the intensity varied from season to season. Some of the crew thought there was a faint septic odor coming from the open water channels. Others thought the area smelled a bit funky but that it was comparable to odors one might expect in a marsh or wetland. Wetlands are prime habitat for sulphur-eating bacteria. Hydrogen sulphide has that familiar rotten egg smell. Methane, however, is odorless, but there are a host of biochemical reactions that take place in wetlands, which is the reason for their efficacy in wastewater treatment. I was pretty indifferent to the smell. I'd gotten used to it.

The wastewater holding tanks were fitted with an alarm to signal when one was full and another needed to be opened. Overflowing sewage was not a likely threat. I checked on the status of the tanks every night, before going to bed and often left valves open so wastewater could flow from one tank to another, without overfilling or leaking.

The author harvesting fodder from the constructed wetlands to feed the goats and chickens inside Biosphere 2.

As my two year stint in Biosphere 2 drew to a close, we counted down the remaining days with more than a little regret. Living inside was a supremely satisfying experience. We were vital parts of that world—the only species that could keep the machinery and environmental technologies going without which life couldn't have continued. On the other hand, we were sustained by the existence of the other life forms inside Biosphere 2. Even inert concrete pillars had played their part, for they had absorbed carbon dioxide. Discovering that eventually helped us solve the mystery of where oxygen had disappeared.

We had watched Biosphere 2 mature. Now there were forty foot trees in the rainforest and savannah, new coral in the ocean, and newborn animals. We biospherians had matured too. We had developed a profound appreciation of how we are completely dependent on the biosphere for our very existence. This was knowledge we now understood at a cellular level because in a small world, this reality is always with you.

We joked that we were actually two years *younger* because of our low-calorie, high protein diet, designed by biospherian Dr. Roy Walford. Our living world was half machine and half organic. As it is for sailors or astronauts, the regular hum of pumps and motors was reassuring. There was also a periodic roar from vacuum pumps sucking up thousands of gallons of water in the ocean, in order to make the waves, which kept our coral reef happy. It sounded like the "old man of the sea." And mechanical sounds merged with noise of crickets and frogs while the air was somehow denser than Earth air and fragrant with the tropical. Indeed, our atmosphere had more carbon dioxide than Earth's and reflected the intensity of the world of life packed into Biosphere 2.

We had adapted to this new world. I found it strange and sometimes difficult to readjust to Earth's environment after I left

Biosphere 2 on September 26th 1993. On the last day, I made my rounds, saying goodbye. We all had been on a journey together. I'd seen my constructed wetland transform all the shit we sent its way. And so I mulled over what I should do next. How could I apply what I had learned. How could I make a difference back in Biosphere 1, planet Earth?

The biospherian crew during the first closure experiment (1991–1993) on the beach next to the ocean after one year inside. From left standing are Abigail Alling, Mark Van Thillo, Mark Nelson, Sally Silverstone and Taber MacCallum; seated in the front row are Roy Walford, Linda Leigh and Jane Poynter.

The author in the beautiful mangrove biome of Biosphere 2 in wetsuit, collecting litter as part of research comparing nutrient cycling to natural systems. The wetland plants were collected from the Everglades National Park in southern Florida and the system replicated the changes of salinity from freshwater species to red mangroves found adjacent to the ocean. What occurs in nature over many miles was compressed to several hundred feet in Biosphere 2. Before the closure started, many doubted mangrove species could grow at 3900 feet elevation in southern Arizona, but the system flourished and plants grew rapidly.

My Love Affair with Wetlands

"When written, shit does not smell."
—Roland Barthes, French philosopher and critic.

It's a Mad, Mad, Mad World

Leaving Biosphere 2 was a serious shocker for me. I had left a world where everything made sense and where it was obvious that every action you took or didn't take (for inaction also has ramifications) made a difference. Now, back in Earth's biosphere, issues were less clear, the voice of one person or group drowned out by millions of people or the immensity of the Earth. I was determined not to become another cog in the giant machine, another consumer, acquiring stuff I really didn't need. Hmm, what to do next?

There was something about wetlands that deeply fascinated me. They are living systems—consisting primarily of green plants and soil microbes. There are great healing properties to be found in gardens, working with and spending time around a living system. E.O. Wilson, the Harvard scientist and ecologist, gave this phenomena a fancy name, biophilia—an innate love for nature. In our increasingly modern, urban and industrial environments, there is less and less of the living around us. A few shrubs and trees as a backdrop, an artificial, fertilized lawn, sprayed with pesticide so no unwanted weeds invade it, and fallen leaves to be removed, using one of those deafening, polluting,

"Loggers look to harvest the last remaining stand of trees in North America," (the trees inside Biosphere 2), by David Fitzsimmons, *Arizona Daily Star,* Tucson, 1992.

gasoline wasting leaf blowers. Consider how much people's thoughts and behaviors might change if they had a constructed wetland right outside, connected to their toilet.

Since the 1960s, scientists have been developing ways to use constructed wetlands for treating and recycling sewage and dealing with other types of pollution. There is a wonderful Far Side® cartoon by Gary Larson which shows a group of frogs and turtles making their ways through a cacti-filled desert, carrying picks and shovels: "We'll put the swamp here!" The caption reads: "Wetland pioneers!" It's almost as easy as that. What makes a constructed wetland? First, you need a lot of water, enough to saturate the soil, then you build a basin with sealed bottom and sides, add gravel or soil and sand, add your sewage or wastewater, some wetland plants with their microbes and you've got an operating system. With a bit of science and engineering you get the equations you need to size the system appropriate to the need. The type of plants to install and degree of cleaning they can achieve is largely dependent on sizing and climate. The warmer and sunnier the better and the less area needed, since plants and microbes have more energy in tropical climes. For every 10 degree Celsius (18 degree Fahrenheit) increase, biological reaction speeds double.

A constructed wetland works in more ways than one. The environment benefits from the creation of an effective wetland system, people benefit from having their sewage treated, the wastewater is put to good use, and a beautiful garden grows.

Constructed wetlands are local systems. Instead of pumping wastewater to a remote sewage treatment plant, it is processed close to its source, often without any machinery. People become much more conscious of the impact they have on water and nutrient cycles; of what happens when they flush the toilet and to where their shower and laundry water goes.

I'd learned that people using septic tank systems are largely unaware that a septic tank is a living system; a bio-digester. It's understandable; microbes are too small to see. It's hard to bond with anaerobic bacteria the way we bond with beautiful plants. And since we don't like to talk about shit, how septic tanks work was probably never fully explained. Modern septic tanks have to be emptied at frequent intervals because people also "conveniently" dispose of unwanted chemicals down the drain: pesticides, herbicides, paint and paint thinners, cleaning fluids, detergents, you name it. People are heedless when it comes to separating food waste from plates and dishes so that goes into the septic tank as well. People are mostly indifferent to how precious a resource fresh, potable water is, and how copious amounts of this valuable water are used to flush everything away and how that much water overloads septic tanks. In earlier times, septic tanks needed to be emptied every ten to twenty years. Nowadays residential septic tanks become clogged with sludge after two years. All those chemicals flushed down the drain kill the bacteria which digest the sludge. And the volume of water swamps the ability of the system to let solids settle so the

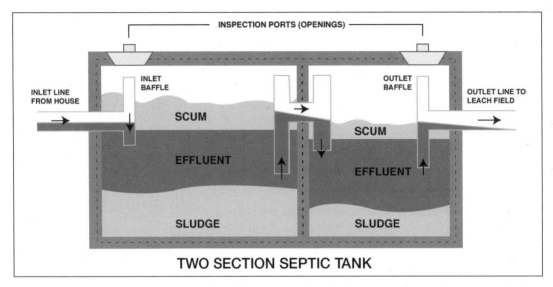

INSPECTION PORTS (OPENINGS)

INLET BAFFLE

OUTLET BAFFLE

INLET LINE FROM HOUSE

OUTLET LINE TO LEACH FIELD

SCUM

SCUM

EFFLUENT

EFFLUENT

SLUDGE

SLUDGE

TWO SECTION SEPTIC TANK

Diagram of a typical septic tank.

microbes can go to work. If a home uses twice the water as before, the sewage stays in the septic tank for only half as long, reducing the time available to separate out the solid waste and digest the sludge. So, more sludge gets through the septic tank, untreated and out to the leach drains. The result is that the leach drain soils become more quickly clogged with organic compounds from the poorly operating septic tank which prevents the soil from absorbing additional wastewater. As a consequence, leach drains too have to be replaced more frequently.

Septic tanks are living systems – anaerobic microbial digesters of sludge (solids) which separates out and settles to the bottom. Kill the microbes with chemicals or overload the system with too much water and it doesn't work. Leach drains get clogged and fail and the septic tank must be pumped out more frequently. A constructed wetland, on the other hand, does a good job at removing organic compounds and suspended solids, which ensures a longer life for the leach drains.

We learned many things from our Russian colleagues during the Biosphere 2 years. Two of the most important were that "you

can count on life to keep
doing what it is biologically
and ecologically evolved to
do and you can count on
machinery breaking down."
A simple truth, perhaps not
obvious to everyone! NASA's
(and most other space agen-
cies') approach to bioregenerative life support systems
takes just the opposite view—
reliance on equipment and a
distrust of the reliability of
organic life systems. Their
aim is to reduce the quantity
and diversity of life they have
to bring to space. A NASA bioengineer once told me: "I want to
control everything, make every individual wheat plant do what I tell
it. Eliminate any ecological interaction between one wheat stem
and another. It's unnecessary!"

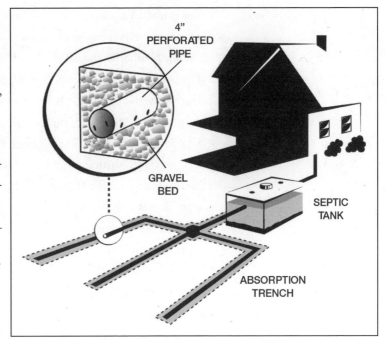

Schematic: A typical "leach drain," designed to dispose of wastewater that comes out of the septic tank by absorption in soil.

The techno-fix! We crave the latest gadgets, gleaming automo-
biles, faster computers or bigger home entertainment centers and
fill our houses with gizmos which are supposed to make us happier,
our lives easier (labor-saving devices) and realize our dreams. But
the technology—and it's not a question of *if*, it's a question of
when—is going to break down. Constructed wetlands, on the other
hand, are reliable, living systems that adapt and change. Of course,
you do need to make sure that you don't flush really toxic stuff
down the drain. If you do that one simple thing, whatever's in your
wastewater will feed the right microbes which will then increase in
number. Living systems last longer than machines and improve as

they evolve. And there's no need for constant monitoring by technicians or the replacement of parts.

So, meditating on all of this, I decided to return to academia, after an absence of twenty-five years, and learn more about constructed wetlands of which I was enamored. My first stop was the University of Arizona at Tucson, 40 miles (64 kilometers) south of the Biosphere 2 site. I knew Lloyd Gay, a professor at the School of Renewable Natural Resources. His wife, Mia, an Estonian, had tried to teach me Russian and had translated for us when Russian scientists visited Biosphere 2. Lloyd had come out to Biosphere 2, had seen our wetlands and was amenable to my studying them as a thesis for a Master's degree.

I suggested to Lloyd we could cut the sewage pipes from the Renewable Natural Resources building and make a demonstration wetland right outside. "Not the way it's done here, I'm afraid!" My two years at U of A were not exactly learning by doing —there were classes, lectures, fieldtrips and study. But there were also compensations. U of A is known for its excellent Arid Lands Studies department as well as water sciences, space sciences and astronomy. Field trips got me out to Arizona's precious few remaining wetlands and riparian areas.

I took foundation courses on water chemistry, soil chemistry, watershed management, hydrology, forest management and comparative desert studies. My timing was auspicious. Interest in constructed wetlands was just getting started. By connecting with like-minded people at the university and around Arizona, I was able to study working examples of different constructed wetlands.

Engineers and ecologists approach things differently. I crossed the dividing line that separates them. In the engineering building, across from the Renewable Natural Resources building, I saw what engineers think about sewage. I felt like a "mole," under deep cover. I heard the awful news: In high-tech sewage treatment facilities, there are....bugs (microbes)! I appreciate that engineers take safety and performance very seriously. The presence of these microbes was so frustrating. I could hear the inner monologue of the civil engineering professor. His words went something like this:

"Yes, you can pour all that concrete, lay all that steel, set up control systems for the pumps, aerators and dryers, but at the heart of the beast, and there's no way around it, you're just going to have to accept that your technology is supporting a mass of bacteria which no one can control and no one fully understands but which is crucial to the whole process. All we engineers can do is provide the right conditions for these invisible critters and hope they do their job, like all the rest of the professionally operated and maintained equipment of which we are so justly proud."

As a participant in a class on environmental water quality, I was able to see a giant sewage plant in operation. In 1995 Tucson was a city of around half a million people and its sewage was pumped to two treatment plants. We went to the one on the north side. When you get close, you understand why these facilities are put in low-rent districts. You can smell them before you see them. We toured the facility and were told it had won awards. It generated much of the power it needed from the methane collected as a by-product of the anaerobic bacteria doing their job.

The first step in treating waste is a giant mesh screen which stops non-biodegradable materials from continuing along in the wastewater stream—metallic objects, plastic wastes, prophylactics, and other trash. Actual organic sewage then continues on to a circular tank where aerators and stirrers go to work. The microbes here are aerobic and need oxygen. Chemicals are now added to precipitate the sludge which is then sent on to a sludge digester. Here, the anaerobes go to work and the methane that is released is collected for power generation. The last step is to add chlorine to disinfect the treated sewage water before it is discharged into the river.

However, in Tucson, there really isn't a "river" out there. Southern Arizona once had rivers but they were long ago diverted to irrigate farms and to supply water to the growing cities. And now, because there is so much sewage processed through Tucson's two treatment plants, the water discharged is regarded by the Environmental Protection Agency (EPA) as a river in and of itself, and therefore must be up to the appropriate standards and quality found in a "natural" river. Thus, there are many inspections and much pressure to operate the facility well so that fewer nutrients and bacteria are sent out with the wastewater, as if the sewage plant were polluting a river. Sad to say, the centralization of shit has created this new type of river which is increasingly found around the urbanized world: "Rivers of Shit."

This idea, although awful, has a long and official pedigree. In a famous French royal edict of 1539, in which severe penalties and loss of property were imposed for disobedience, the citizens were warned: "We forbid all emptying or tossing out into the streets and squares....of refuse, offals or putrefactions, as well as all waters whatever their nature, and we command you to delay and retain any and all stagnant and sullied waters and urines within the confines of your homes. We enjoin you to then carry these and promptly empty them into the stream and give them chase with a bucketful of clean water to hasten their course."

The spread of indoor plumbing has vastly multiplied that bucketful of clean water, so that whole Amazon Rivers of freshwater, now contaminated with shit, are discharged into those once idyllic "streams." To move a ton of humanure requires more than a *thousand tons* of water. Each person using an old-fashioned flush toilet may be using over 10,000 gallons (40,000 liters) per year to flush their

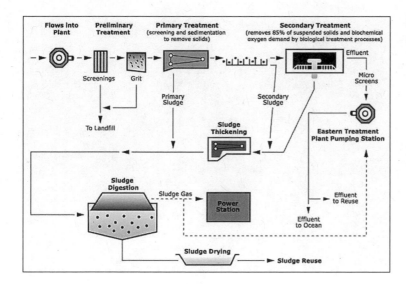

Layout of treatment stages in a conventional sewage plant.

shit away, first to the sewage treatment plants and then eventually to "Shit Creek," the dark and usually ignored side of our now virtually unquestioned modern convenience, indoor plumbing.

The sewage plant shown on the next page is very similar to the one in Tucson, Arizona. Wastewater first flows through coarse screens that filter out large objects and other non-organic material, then proceeds through finer screens and sedimentation filters which remove organic solids (sludge). Secondary treatment usually involves aerated mixing tanks which enable microbial digestion and further separation out of solid waste (secondary sludge). The sludge is now thicker and is sent on to a sludge digester where methane emerges and is used in power generation. The remaining sludge is dried and sent away for reuse or disposal. Treated wastewater is disinfected before disposal in a nearby river, lake or ocean. In rare cases, some wastewater is used on parklands or for golf course irrigation.

At that sewage plant I was introduced to the concept of peak load. My visit took place in mid-February, at the end of the American football season, so peak load was on the manager's mind. He

A modern sewage plant: notice the circular pools, which aerate and separate out sewage sludge, during the secondary phase of treatment.

said that in earlier times it was uncertain when the "peak" would happen but for the past few decades it had been predictable. Peak load in Tucson was half an hour after the halftime intermission of the Super Bowl—when everyone heads to the toilet.

Sewage plants vary in the extent to which they attempt to microbially digest sludge, in order to produce secondary effluent that meets regulatory standards. The standards for city sewage plants call for under 30 mg/liter of organic matter, as measured by the water's demand for oxygen and 30 mg/liter of total suspended solids. Sewage plants usually also introduce a final disinfection step, to kill disease-causing bacteria, before the wastewater is discharged.

Disinfection is usually done with chlorine (because it is the cheapest method) or with ozone or ultraviolet light, all of which kill bacteria, viruses, and parasites. Chlorine has a downside though—it combines with other elements to produce carcinogenic compounds that adversely affect aquatic and marine ecosystems. The need to remove nutrients from wastewater has led to "advanced" sewage treatment or tertiary treatment, whereby nitrogen and phosphorus are brought down to below certain levels, (often 10 mg/liter nitrogen and 1mg/liter phosphorus).

Capital costs for this type of system are high. US government outlay on wastewater treatment infrastructure comes just after highway construction spending. In second stage sewage treatment, there is no specific requirement to reduce nutrients, as a result of which discharges from these plants cause eutrophication (excessive nutrients and consequent depletion of oxygen) of natural waters. In recent years more and more communities are building advanced treatment systems, which produce tertiary level wastewater, where nitrogen and phosphorus are further reduced. But this only changes the *form* of the problematic "waste." It does not eliminate it. The more advanced the treatment and the cleaner the treated wastewater, the more sludge that is produced, since it contains the materials filtered out.

This sludge could readily be reused if it only contained wastewater from people. But because urban sewage is a mix of commercial and industrial waste and far more harmless residential waste, the sludge contains heavy metals and synthetic chemicals. No one knows exactly what goes down all our drains and on to the sewage plant. Some experts think Americans dump millions of gallons of used motor oil each year. Every city has a different mix because each has different industries and there are around a hundred thousand different synthetic and organic chemicals used by industry (along with maybe a thousand new compounds invented every year). Little is known about the long-term effects of exposure to these potent synthetic chemical mixes. And, of course, it is expensive to analyze all the constituents in this mélange called urban sewage plant sludge. Disposal options have become very limited. Sludge used to be incinerated, buried in landfills or dumped in the ocean. But clean air regulations long ago ruled out incineration; evidence of massive water table pollution made burial in landfills illegal; and in the early 1990s, the United States stopped ocean dumping. Land application is the only real option, but this is controversial because of the unknown effects of applying sludge to crops. The EPA (US Environmental Protection Agency) has euphemistically renamed sludge "beneficial bio-solids" and it is often offered free of charge to farmers, as fertilizer. However, many farmers remain reluctant to accept it because of the potential adverse effect on soil and crops. Regulations designed to determine how sludge may be used on farmland are evolving but provoke much argument.

The more I learned about conventional methods of sewage treatment, the more attractive the alternatives looked.

The Mighty Cottonwood

The subject of my master's thesis was zero discharge constructed wetlands, planted with fast-growing poplar and willow trees. Zero discharge means systems that consume *all* the wastewater fed them. There would be no concern about the quality of the discharge water because all the water would be used to grow trees. Plants are like swamp coolers; they use a great deal of water to keep themselves cool, a process called transpiration. There is also evaporation going on. It's difficult to measure either of these directly, and their combined effect, evapotranspiration, has spawned a scientific cottage industry of equations and equipment. In dry climates, this water usage can be prodigious: It can take an estimated twenty-five to fifty tons of water to produce one ton of corn, depending on the aridity of the climate. Lloyd Gay, my adviser, was one of the world's experts in this field. He'd spent a good part of his life wiring up towers and trees to measure where all the water goes. This accounted for his interest in my thesis, using trees to "get rid" of wastewater in a productive way. Poplar and willow trees are ideal for wetlands since they can pump air to their roots even in water saturated soil. These trees are fast-growing;

provide habitat and food for animals and the wood they produce is a harvestable commodity. They can be coppiced (cut back), every few years, since with their robust root systems, they quickly grow again.

Back to the Garden

I was also very interested in subsurface flow wetlands (where the wastewater is kept a few inches below the surface of a bed of gravel) because they offer some advantages over surface flow wetlands. Since sewage is never present on the surface, there is neither odor nor the risk of disease through accidental contact. Also, mosquitoes can't breed since there is no standing water. Furthermore, the intensity of treatment is greater because the entire surface area of gravel gets colonized by bacteria. Thus, a smaller area, usually around one-fifth the size, is needed to achieve the same level of treatment that a surface flow wetland would require. The downside is that such a system is more expensive.

Recalling wading in front of tourists while doing my duties at Biosphere 2, the idea of a layer of gravel between me and the shit had great appeal! In some cases, surface flow wetlands are the better choice, as for example, when the wastewater has high levels of organic compounds and suspended solids since all that material would clog the gravel used in subsurface wetlands. This is the case at animal farms raising chickens, cows or pigs, or at food processing plants. So, surface flow wetlands are becoming popular as a cost-effective and natural way of treating and recycling farm wastewater. In many cases, constructed wetlands can also provide additional benefit since the plants used can provide fodder, which is fed back to the animals. With some level of disinfection, wetlands can also improve water quality sufficiently so that the treated water can be used for flushing toilets, thereby reducing overall fresh water consumption. Wastewater can be used for irrigation or compost-making, thus getting remaining nu-

An open water or free water surface constructed wetland for wastewater treatment. Most of the biological activity occurs in the top layer of the soil, the stems of the plants and in the water itself. Frequently there is no liner under this type of constructed wetland and there may be water which percolates into the soil or groundwater, depending on the permeability of the soil.

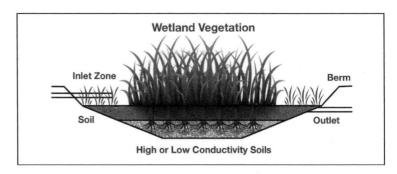

trients back to soils and plants. In a surface flow wetland, treatment areas are principally the top layers of the soil, the stems of the plants, colonized by microbes, and the microbes in the exposed wastewater.

Fortunately, I learned about an actual system operating in Tucson using poplar and willow trees, from Martin Karpiscak, a research scientist with the Office of Arid Lands Studies. At the older of Tucson's sewage plants, a small fraction of wastewater was being sent through a series of parallel constructed wetlands. Some were filled with floating plants but several had cottonwood and willow trees. These trees are often found along riverbanks. Here they stood, in two to three feet of wastewater soaked gravel. The wetland liner was a black rubber-like geomembrane. Over time, cottonwoods get so tall they are liable to fall over in high winds, so they were pruned regularly. Red-tailed hawks had arrived and made their home in the grove. I realized that constructed wetlands are a natural analog to

A subsurface flow constructed wetland. Here the gravel bed is 2 inches (5 cm) above the level of the wastewater. These systems typically have a water-tight liner of geomembrane such as EPDM, compacted clay, or (in countries with inexpensive labor) concrete.

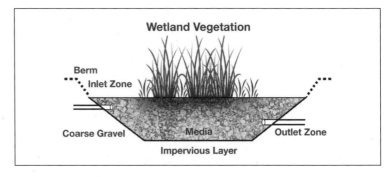

View of cottonwood, willow and other species growing at the Constructed Ecosystem Research Facility in Tucson, Arizona (sponsored by Pima County Wastewater Management Department). The trees are growing in lined raceways in 5 feet (1.5m) of gravel and larger stone flooded to a depth of 3 feet (1m) with either potable water or wastewater. The growth rate was several feet (1m) a year and in some instances as much as 10 feet (3m) a year. (Photo courtesy of the Office of Arid Land Studies, University of Arizona.)

hydroponic systems. No soil is actually needed. Instead of having to buy fertilizer to give the trees the nutrients they require, animal and human waste can do a fine job, for free.

At the end of 1995, I completed my masters of science in watershed management at the School of Renewable Natural Resources. I now made plans to join the creative ferment that was always brewing around the legendary scientist, Howard T. Odum, at the University of Florida, Gainesville.

Gainsville, Florida–
Studies with Masters of Systems Ecology

> *"And do not spread the compost on*
> *the weeds to make them ranker."*
> –William Shakespeare: *Hamlet*, Act III, Scene IV.

I met H.T. Odum when I was at Biosphere 2. He and his brother Gene were invited to Arizona to participate in our lecture series and research programs. The Odum brothers had revolutionized ecology and were acclaimed as the founders of "systems ecology,"

H.T. Odum and the author examine vegetation inside Biosphere 2 in 1994 after the completion of the first closure experiment. Prof. Odum was especially excited about the demonstration of ecological self-organization which had occurred in all its biomes since the start of the experiment.

the study of ecological interactions from a systems perspective. H.T. was also known as the father of ecological engineering since he had been one of the first to advocate the use of natural mechanisms to solve environmental problems. Constructed wetlands were a great example of this approach.

H.T. Odum was professor of Environmental Engineering Sciences at the University of Florida, Gainesville, where he had set up a multidisciplinary Center for Wetlands to promote understanding of natural wetlands, the restoration of damaged wetlands and the use of constructed wetlands in sewage treatment. He was a quintessential holistic thinker. In his Thursday afternoon seminars any topic was fair game but H.T. insisted that a systems diagram be drawn first, so that everyone got an overview of what was being discussed. H.T. was feisty and loved to upset the status quo, especially when that status quo was built upon poor logic or was being cited to rationalize social injustice. He was a catalyst for independent thinking since he'd make mincemeat of anyone foolish enough to spout jargon or rote recitations in his presence.

There are two distinct approaches to the scientific study of natural phenomena: one is the reductionist and the other, the holistic. H.T. explained the animosity that some reductionists feel toward holistic science. These reductionist scientists (who try to understand nature by working with the smallest, most fundamental components) think they are doing the only "real" science. They don't accept that science can also include the systems analysis people who work at higher levels, seeking to understand nature as a complex, hierarchical whole, instead of just focusing on the smallest and simplest components.

Odum was a pioneer in wetland science. In the 1970s, his Center for Wetlands had experimented with sewage treatment using natural cypress swamps found in Florida and the southeast US. The work was successful, but ironically, greater environmental protection of natural wetlands has restricted their use in sewage treatment. So Odum turned to alternative approaches, including large-scale bioengineering of wetlands and restoration of areas damaged by phosphate mining.

Up to My Waist in It

Most of Florida is flat as a pancake and it rains a lot. There aren't many slopes to drain water away. When I first arrived in Gainesville, I was amazed to hear people talk about high elevations, when they meant 30-60 feet (10-20 meters)! In Florida, six inches elevation in a wetland can make a large difference in determining what sort of vegetation grows. I listened to discussions about "micro-typography." Any kind of dip in the ground and you'd have yourself a wetland.

One spring there was a spate of accidents with motorists hitting alligators lying by the side of the road. Heavy rain had extended wetlands right to the edge of the highways. Florida is mostly limestone, a soft rock which easily wears away and forms the karst topography associated with cave formation. Limestone erosion also leads to the formation of sinkholes that sometimes break through to the surface, unexpectedly. I saw photos of a house and some cars in Gainesville that had simply disappeared one night, down into the black depths of a sinkhole.

Since rain combined with carbon dioxide in the air forms carbonic acid, which dissolves limestone, the landscape gets eaten away and wetlands take shape in the lower ground. Sometimes a lake can evolve into a wetland, especially if a sinkhole opens and surface water goes down into it. Payne's Prairie, near Gainesville, had once been a lake, but in living memory had morphed into a beautiful wetland.

Although it's in Florida, Gainesville can get cold. It can freeze in the winter. The citrus industry has retreated south in the past few decades because of severe frosts. Scientists say this is due to so many wetlands being drained for agriculture or real estate development. The moderating effect wetlands have on climate has been reduced.

It was a rite of passage then, on a freezing day in February 1996, to be hopelessly mired in the muck of a Florida salt marsh flat. It was my first Wetland Ecology field trip. Clay Montague, one of our professors, chortled with glee. I learned on a field trip that Clay had studied mangrove fiddler crabs for his PhD. He could wax lyrical about the synchronized way males wave their large claws as a mating signal. It was so boggy that some students crawled out on their hands and knees; others used plastic buckets as a third leg to prevent sinking further into the quicksand-like marsh soils. Others clung to each other in twos and threes, weaving crazily across the salt marsh seeking higher, firmer ground. Occasionally, you'd sink to your ankles if you were lucky, and to your knees or waist if not. By the time we made it out, the mud had freeze-dried on our clothes, making our jeans ten times heavier. Anyone who spends their time in mangrove wetlands and salt marshes has clearly learned to love being in the muck and mud.

Walking on Water

Welcome to the wonders of wetlands. Over the next few years I would delight in so many. There are cypress domes, dominated by trees with bony projections (knees) which everyone suspects has something to do with aerating the cypress's roots. The dome shape arises because the trees in the center, where more peat is deposited, grow taller than outlying ones. Another type of wetland is the salt marsh, usually by the coast, dominated by salt grasses and rushes but hardly any trees. Mangrove wetlands come in many forms. There are red mangroves with their entangling prop roots, black mangroves with their pneumatophores (breathing roots) poking up from the ground like spines on a porcupine; white mangroves, buttonwood trees and wetland ferns that grow to nine feet high (3 meters). Lakes and rivers often have fringing wetlands shallow enough for plants to take root in the bottom soil and emerge through the water. Hence, "emergent wetland plants," as opposed to water lily, water lettuce and hyacinth, wetland plants which float.

Perhaps the most amazing are the floating wetlands. We went to study one at Orange Lake. We went out by boat. The wetland was only a few acres and it had a peculiar spongy quality. It was suspended in ten feet of water. There wasn't really any soil. Wetland plants deposit peat (which forms peat bogs, another type of wetland from where we get peat moss) and their root systems release enough gas to keep the island "floating." Floating wetlands can be tiny or can cover large areas but they are light enough so that strong winds can blow them across lakes, moving them around. We traipsed around our floating wetland, measuring plant cover and biodiversity. Occasionally, we would fall through to the bottom and get soaked. The scene was surreal. We were literally walking on water.

Wetlands as Laboratories

K. Ramesh Reddy is an expert on wetland soils. I was fortunate to have him on my dissertation committee. His course on the biogeochemistry of wetland soils was a requirement and fundamental to understanding how wetlands work. Reddy introduced us to the world of anaerobic microbes and how life can survive without oxygen or sunlight. Photosynthesis may be going on at the surface but down below is an environment akin to an earlier time on Earth, before oxygen (once toxic for living organisms) was released into the air in such huge quantities that the atmosphere was completely transformed.

I began to understand why wetlands are such powerful cleansing agents. They can break down seriously toxic man-made chemical compounds. A wetland's anaerobic environment hosts some of the toughest microbes Nature has created. Down in the anoxic wetlands, the first substances to be metabolized are nitrogen and phosphorus compounds, and this process releases a great deal of energy. The next to be broken down are the tougher ones—sulfur and carbon compounds. By the time methane-consuming microbes get to work, there is less energy available, so not all the carbon in methane (CH_4) gets consumed. The residual carbon is deposited in wetland soils which are characterized by being non-oxidizing ("reducing") environments. The carbon is deposited rather than burned up as it would be in normal (aerated) soils. Thus peat is formed and it is created in every type of wetland. The more anaerobic, the more organic carbon is deposited.

Ramesh Reddy is a naturally aristocratic and soft-spoken man. His laboratory is also involved in research on the problems of the Everglades, swamps which cover a large portion of south Florida. Part of the original Everglades was drained and converted to agriculture, primarily for sugarcane. Much of the water flowing through the Everglades has been diverted to supply the growing cities in the region, and redirected into reservoirs for flood control and irrigation. "The River of Grass," as the Everglades was once called, now has its water regulated by an elaborate series of dams and other controls. The number of wading birds in the Everglades has been severely reduced. It is said that the Everglades is the only US National Park that is threatened with *total* ecological destruction because water flow is so crucial to wetlands. When wetland soils are drained and then exposed to oxygen, the peat literally burns up (oxidizes). The Everglades' sugar cane fields lose inches of soil within a few years and may soon become useless for agriculture. Further exacerbating the threat, all the fertilizer and pesticide runoff from the sugar cane fields flows south, through the Everglades, promoting the growth of invasive weeds which compete with native vegetation for nutrients. It all is so shortsighted and everyone knows it, but profits and a "business as usual" approach takes precedence. The year I arrived a campaign to place a one cent per pound tax on sugar, to raise funds for the restoration of the Everglades, was defeated in a Florida referendum.

Wetlands are complex laboratories. The interaction between well aerated layers near the surface and anoxic ones below varies according to the season and soil and vegetation. Many nutrient cycles have both aerobic and anaerobic components. The aerobic process is faster and releases more energy but anaerobic processes dominate because wetlands are, for at least a portion of the year, saturated with water.

We learned how to distinguish wetland soils and understand the myriad physical, chemical and biological reactions occurring simultaneously. Wetland plants have developed special devices to send air to their root systems. Some of that air leaks out and forms a thin oxygenated zone around the roots, which is how aerobic processes occur as well as anaerobic ones.

Identifying the Players

Studies in wetland ecology include plenty of fieldwork. It was like a crash course in botany, learning the characteristics and names of some hundreds of plants. It was overwhelming, at first. I began to really see the enormous biodiversity in wetlands. There was everything from ferns to floaters (non-rooted plants that cover the surface of open water), bushes and towering trees; vines, reeds, cattails and spectacular flowers. I also started to see how merely because a plant could survive in a wetland didn't mean that it was restricted to that habitat. Some plants can grow in "normal" aerated soil or in wetlands. Mangroves and other halophytes (salt-loving plants) grow better the more fresh water they receive. But because these halophytes can survive the salinity found in many wetlands, they dominate in coastal and salt marsh wetland areas where most other plants, of course, cannot endure.

Wetlands as Buffers

Hanging out with these ecological systems experts, I too began to see everything as systems. Constructed wetlands are man-made systems that emulate natural ones. Natural wetlands are the interface between the land and water, whether coastal wetlands, floodplain wetlands or wetlands at the edge of lakes and ponds. They are rightly referred to as the kidneys of the planet. They are the systems that can detoxify wastewater and transform it.

The Swamp Doctor

H.T. Odum's circle weren't just theorists. These people went out and did practical projects too. Mark T. Brown, a professor in the systems ecology group, was nicknamed "The Swamp Doctor." He could diagnose problems in natural wetlands, which were usually the result of human activity, and recommend solutions. Dr. Brown also led a group designing constructed wetlands to be used in the treatment of effluent. Some of these constructed wetlands are huge. There are two in the Orlando area that cover over a thousand acres each, which serve a number of functions. First: replacement of previously lost natural wetlands. It's now law in Florida and many other states that if a development destroys wetlands, an equivalent area of *new* wetland must be created. Second: wetlands provide further cleaning of wastewater that has passed through a conventional sewage plant. The EPA has sometimes mandated the installation of a constructed wetland in communities which already have a conventional sewage plant, in order to further improve the sewage treatment and water purification. The third service: recharging the aquifer. Florida, previously abundantly supplied with fresh water, now faces water shortages because of massive increases in consumption from increasing population and the spread of agriculture, industry and tourism. By sending partially cleaned water through a surface flow constructed wetland, there is the opportunity not only to create a beautiful area but to allow the water time to become even cleaner and then percolate down and recharge the aquifer. This solution I first saw in Florida deeply impressed me. Unlike coastal sewage plant managers around the world, the people running this one recognized that freshwater, even treated wastewater, is a scarce and precious resource and it would be tragic if it were simply discharged into the ocean. Large wetlands are beautiful. It amused me to realize on one field trip that few people enjoying this "recreational area" were aware that the constructed wetlands they biked and walked around were also treating water containing human waste. Wading birds like herons and egrets enjoyed these wetlands as well, all while they received and processed millions of gallons of wastewater every day.

Typical cypress swamp found throughout the Southeast of the United States. Odum and the Center for Wetland had demonstrated the ability of this type of wetland with its peat soils to absorb and purify sewage. Legal protection and restrictions on the use of natural wetlands has accelerated the development of constructed wetlands for the same purpose.

The reconstruction of wetlands in the central Florida phosphate mining region was also very exciting. We stopped by to watch giant excavators at work, stripping away thirty feet (10m) of the "overburden" (natural soil) to reach the phosphate layer. One machine can demolish an acre per hour and several of them were working around-the-clock, destroying everything in the way of extracting the deep layer of phosphate minerals. After all the phosphate is extracted, by law the environment must be restored as closely as possible to its previous condition. Enter the Swamp Doctor and his team. They have not only restored wetlands but also forests, grasslands and other biomes that covered the area before the excavators wrought their havoc.

We toured some of these constructed wetlands, now teeming with plants and hosting a wide variety of birds and animals. It was hard to realize that the land we were exploring was once a bleak, post mining, wasteland. Mark Brown talked about some of the tricks of the trade, such as bringing in dead trees or tall posts to provide roosting for birds. Birds excrete seeds they have eaten at other wetlands, which helps replenish the diversity of plant life essential for restoring a wetland.

Aerial view of the town of Akumal, along the Yucatan coast of southeast Mexico.
Rapidly developing, this coastline south of Cancun is being called "The Riviera Maya."

5
Magic in Mexico

"Ah, Love! Could you and I with Fate conspire
To grasp this sorry Scheme of Things entire,
Would not we shatter it to bits—and then
Re-mold it nearer to the Heart's Desire!"

—Omar Khayyam: *The Rubaiyat* (Fitzgerald translation).

THE YUCATAN IS AN EXTRAORDINARY LAND. It is home to the Maya people whose ancient, pyramid-shaped monuments emanate mystery and grandeur even in ruins. Ecologically, the Yucatan is similar to southern Florida—a virtually flat sheet of limestone. There are no rivers. Fresh water is obtained through *cenotes* (water-holes). A magnificent coral reef fringes the Yucatan and extends down the east coast of Mexico, to Belize and further south. Inland, there is one of the greatest remaining areas of intact tropical forest.

In the 1970s the city of Cancun was a small fishing village. The Mexican government then decided to develop it into an international tourist resort. By the mid-1990s, it had a permanent population of over 300,000 with two to three million visitors a year. Akumal is 70 miles (110 kilometers) south of Cancun and is a charming hamlet. It has a couple of half-moon shaped beaches and the Yal-ku Lagoon, inhabited by a dazzling array of tropical fish. It is said to be the site where a shipwrecked Spanish sailor became so possessed by the charm of the Maya and their land that he refused to be "rescued." He chose to remain and married a beautiful Mayan woman. He was the first European to set foot in the Yucatan.[2]

The corals for Biosphere 2's ocean had mostly been collected off the coast of Akumal. The team designing the artificial ocean had met Gonzalo Arcila, a native of Merida, the Yucatan state capital. Gonzalo had helped with the collection of live coral to be sent the thousand miles to the Arizona mountain desert site of Biosphere 2. The Mexican government had issued special permits to allow this. Biosphere 2's million-gallon (4 million liter) ocean with its coral reef of nearly fifty different species still ranks as the world's largest living man-made coral reef. It is also the farthest north and highest in elevation (4000 feet, 1000 meters, rather than sea level) of any other reef on the planet. In gratitude, we Biosphere 2 principals promised that we would return to help solve the sewage problem along the Yucatan coast.

Abigail (Gaie) Alling, a marine biologist and one of my Biosphere 2 friends, had started a non-profit organization called the Planetary Coral Reef Foundation (or PCRF, a division of the Biosphere Foundation) to work on protection and monitoring of the world's coral reefs. Coral reefs are in serious trouble but, because they are underwater and thus out of sight, their dire situation is not as appreciated as that of the tropical rainforests. Coral reefs are not even well mapped. We do not even roughly know how much area they cover on Earth but they are a critical part of our global biosphere. Reefs are the marine equivalent of rainforests, highly productive and containing an astonishing diversity of life. Many factors are contributing to coral reef decline, but a leading one is pollution from human effluent. Coral reefs, like rainforests, thrive in low-nutrient environments. When extra nutrients are present, algae thrives and overcomes the coral, depriving it of sunlight for photosynthesis. There are, of course, other problems caused by inadequately treated sewage, including increased turbidity of coastal water, loss of dissolved oxygen, adverse side effects due to organic compounds

[2] Juan-Navarro Santiago and Theodore Robert Young (Editors). *A Twice-told Tale: Reinventing the Encounter in Iberian/Iberian American Literature and Film,* London, 2001, pp 137-139.

in sewage and damage caused by chlorine used to disinfect sewage before it reaches the ocean.

The PCRF received some modest grants: $20,000 for a field station and $14,500 to demonstrate a constructed wetland. We were confident that we could do a lot with little, as at the Institute of Ecotechnics, where our goal was by being resourceful and doing a lot of the work ourselves, we could make one dollar do what would normally require ten. $7000 of grant money went to the Center for Wetlands back at the University of Florida, to procure their expertise, and that garnered a three year tuition waiver and a research assistant position for me.

Dr. Mark Brown, our Swamp Doc, and I made our first trip to the Yucatan in March 1996. Unlike the tourists who crowd Cancun's hotels, we were on a mission—sleuths on the trail of shit. Where is it going? What effect is it having? How can we improve things?

It wasn't hard to get clues. The inner lagoon at Cancun smelled like an open septic tank. We learned that many of the hotels simply sent their shit straight into the lagoon and the overburdened city sewage treatment plant was also discharging there. "Another margarita, *señor*?" "Make that a double—so I can ignore the smell."

The Disappearing Act

The main Akumal beach is open to the public and is enjoyed by Mayans, Mexicans and gringos, alike. Beachfront houses and hotels at Akumal were opulent by Mexican standards. With the cost of living low and labor cheap, you could afford to build in a grand style. But this grandiose sense did not extend to treating shit. As Mark Brown and I walked the main street of Akumal, we could smell our way along. Houses either had unlined septic tanks or an unlined hole (a "cesspool") near a septic tank through which effluent was discharged. Many of the condominiums and smaller

Prof. Mark T. Brown examining vegetation near the Yucatan coast, Mexico.

hotels caused another headache. They poured their septic tank discharge (septage) into what are called "deep injection wells." I had been investigating these when I was researching coral reef problems in the Florida Keys. There, the deep injection wells typically went down 1500-2000 feet (450-650 m). But deep injection wells do not offer a good sewage disposal solution because the Keys, like the Yucatan, are made of limestone (which is very porous). Researchers from the United States Geological Survey had shown that effluent in these deep injection wells was surfacing less than a mile from shore. Adding to this problem is that, like fresh water, wastewater is lighter than the salt water so it quickly rises to the surface, which puts the shit right in the middle of the coral reef. In the Yucatan "deep-injection" means drilling holes just 150 feet (50m) deep. Attempting to make shit disappear in this way is a grand illusion. One local at a cocktail party in Akumal told us: "We don't like swimming or diving in the reef anymore. We know it's dirty and you hear more and more stories of encounters with 'brown trout.'" Another euphemism to add to my growing inventory of ways to *not* say shit.

Over the years, on many trips to so-called developing countries, I have become accustomed to the idea that someone builds a septic tank but doesn't seal it by means of a concrete floor. Think of the convenience. No need to pump out accumulating solids, no need to build a leach drain. The whole mess just goes straight into the ground. When that ground is porous limestone, the shit goes straight out to the ocean and the coral reef, as well.

If a septic tank system was actually sealed, another creative local practice was to call for a pump truck to take the wastewater away. Supposedly, the pump truck drives to the nearest sewage treatment facility; in Akumal's case, at the nearby town of Playa del Carmen. However, it was common knowledge that what often happened was that the truck driver saved fuel, wear-and-tear and the sewage plant fee, by driving a short distance away, finding a convenient track into the mangrove swamp or the rainforest, and dumping the load right there.

There were plenty of citizens in Akumal and elsewhere along the coast who were concerned about this state of affairs. The area was trying to promote ecotourism, which was very much in vogue, and snorkeling and scuba diving were major tourist pastimes.

A center for ecology (Centro Ecologico Akumal, CEA) had been started in Akumal, to protect the nesting places of sea turtles, the mangrove swamps and wetlands and promote recycling and sustainability.

The Swamp Doc and I also toured some of the area's wetlands to take inventory of plant resources available for constructed wetlands, and to gain more understanding of the local ecology. The Yucatan abounds in wetlands, perhaps more so than northern Florida. The coastal topography consists of rocky headlands or sandy beaches by the ocean, sand dunes off the beach, a mangrove area farther back followed by tropical forest still further inland. When it rained, low elevation areas morphed into wetland. Small wetlands could be found along the highways or beside dirt tracks.

Gonzalo Arcila, founder of the Akumal Dive Shop, shows Mark Brown (in cap) and the author (in large straw hat) some of the wetland biodiversity of the Yucatan coast.

We decided there were plenty of good candidates for the constructed wetland plants. The CEA donated some land for the PCRF's field station and land for two demonstration wetlands. One would service the PCRF's building and some nearby offices, and the other would be for the dormitories where visiting scientists and CEA volunteers stayed. The donated land had recently been laid waste by a hurricane. It was a leap of faith to imagine a constructed wetland there.

Between a Rock and a Hard Place

My design for the two wetlands called for one measuring 22 feet (7m) on a side, covering 500 square feet (to serve sixteen people) and the other, 29 feet by 29 feet (9m x 9m) or 850 square feet

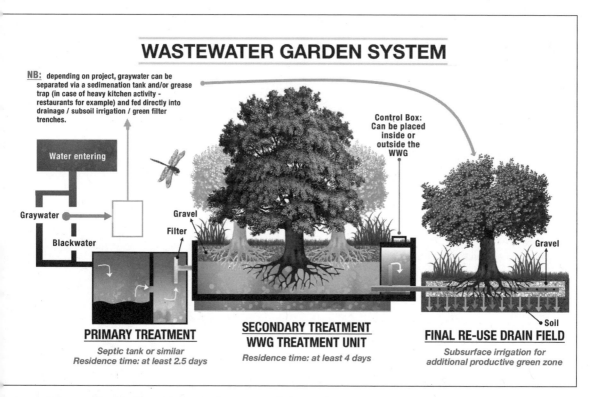

WASTEWATER GARDEN SYSTEM

NB: depending on project, graywater can be separated via a sedimenation tank and/or grease trap (in case of heavy kitchen activity - restaurants for example) and fed directly into drainage / subsoil irrigation / green filter trenches.

Water entering

Graywater

Blackwater

Gravel Filter

Control Box: Can be placed inside or outside the WWG

Gravel

Soil

PRIMARY TREATMENT
Septic tank or similar
Residence time: at least 2.5 days

SECONDARY TREATMENT
WWG TREATMENT UNIT
Residence time: at least 4 days

FINAL RE-USE DRAIN FIELD
Subsurface irrigation for additional productive green zone

Schematic for the Wastewater Garden System (WWG)*

*Wastewater Gardens® is a registered trademark of the Biosphere Foundation.

(for twenty-four residents). Each wetland would have two cells or compartments and would be fed by gravity flow from nearby septic tanks. The wetlands would be subsurface flow. They would be made of concrete which was cheaper than importing a geomembrane liner. Our project would employ locals. Limestone gravel was in plentiful supply and was not expensive. Wetland plants were obviously available locally, so the only nonlocal materials needed were small amounts of PVC pipe, some cement, reinforcing steel and septic tank filters.

In the previous chapter we discussed subsurface flow wetlands. There are two types. One is called horizontal flow, where the wastewater simply fills each compartment and conditions are mainly anaerobic. The other, called vertical flow, needs a pump or siphon

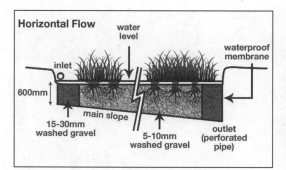

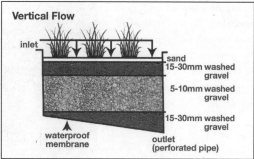

system to fill at least one of the wetland cells. Inflowing wastewater is routed through a distributed network of perforated pipes laid out on the surface of the wetland gravel so the wastewater seeps down through the gravel to another compartment. This wetland is much more aerobic, which increases its ability to reduce nitrogen.

I decided on a horizontal flow type of wetland because, in our experience, the more expensive and technically-demanding the system is, the less likely it will be well maintained. With no pumps or chemicals needed, the system would be less costly and easier to manage. My guiding philosophy is to not simply to frame the problem as "treating the wastewater." The objective is to *use* the wastewater to nourish plants, trees and other new life. Our system included not just the Wastewater Garden but made further use of treated wastewater to irrigate suitable trees, shrubs and flowers. Our goal was zero discharge which meant complete re-use. In locations where this is impossible, or in otherwise extremely ecologically sensitive areas—despite their added expense and maintenance requirements—vertical flow subsurface wetlands are an appropriate choice.

Gaie Alling, Mark Van Thillo (Chief Technical Officer at the PCRF), and I met up in Akumal in August 1996 to organize and supervise the installation. Originally from Belgium, Van Thillo was also a member of the first Biosphere 2 crew. I managed the plant

Two basic types of subsurface flow constructed wetlands. On the left, horizontal flow, where the wastewater fills up the wetland compartment and overflows when more water enters. On the right, a vertical flow system showing how incoming wastewater is pumped to a network of pipes on the surface, and then the wastewater flows downwards. Courtesy of Waterways and Wetland by Alan Brooks and Elizabeth Agate, BTCV Handbooks.

During excavation of the Wastewater Gardens in Akumal, we discovered that groundwater was so close to the surface during the rainy season that we had to pump out the water to construct the systems.

Gonzalo Arcila managed the business of building Wastewater Gardens in Mexico, and his partner, Ingrid Datica supervised the plant collections.

collections while he led the construction. Gaie handled the finances and local politics. Gonzalo Arcila, who had helped with the Biosphere 2 coral reef collections at Akumal where he had started the Akumal Dive Shop, and his colleague, Ingrid Datica, a Venezuelan who had settled in Akumal, pitched in.

The first shock was not long in coming. I had pictured our group joining the local construction team, grabbing shovels to dig out the wetland site. After all, it only needed to be about 3 feet (1m) deep to allow gravity flow of the wastewater. A miscalculation! Unless you're standing on sand at the beach, limestone rock is only a few inches below the ground. It is very hard. The backhoe we'd brought in just scratched the rock surface. We had to rent a couple of jackhammers to excavate both constructed wetland spaces.

The next surprise was that we hit groundwater less than 2 feet down. It was the rainy season at the time. Rainwater from the inland tropical forest was flowing underground through the limestone only to meet saltwater from the ocean, creating a briny subsurface lake. It shouldn't have been so surprising that groundwater was so close to the surface. One wetland site was barely 25 feet (8m) from a mangrove swamp. Our original budget was washed away. Happily, the CEA came aboard, providing alternative land and contributing funds toward the cost of septic tanks and constructed wetland sites.

Plant Candidates

Meanwhile the hunt for wetland plants went on. We had decided, since these would be experimental wetlands, that we would plant as wide a variety of plants as possible, then let them self-organize. The plant candidates that survived would demonstrate their suitability for the local conditions. I worked with Ingrid Datica and got some help from a Hungarian friend, Reka Komaromi, who was developing a farm in Belize. We went into the jungle and transplanted native grasses, shrubs and trees from perhaps a dozen different areas. I was forever leaning out of car windows, spotting plants we hadn't seen before nor yet collected.

The locals knew which plants could live in water and which could not. I was initially skeptical. They pointed out many plants that, as far as I could tell, had never been identified as wetland-tolerant, for example, the flowering oleander. But I saw that there were indeed oleanders growing in wetland areas, in the nearby tropical rainforest. These Maya had spent their lives in this region. Time proved their evaluations to be totally accurate.

After three weeks of construction, we were ready to fill the wetlands to make sure they were watertight. Had we mentioned to the construction crew that the concrete had to hold water? There were leaks at the control boxes, where the collector pipes at the end of each wetland compartment fed the water into a standpipe. The operation of the standpipe was simple enough. When more wastewater came in, either from the septic tanks or from rainfall, the standpipe would send the overflow from the first to the second compartment

One of the two initial Wastewater Garden systems in Akumal, Mexico. The concrete liner completed, the perforated collection pipes at the end of each treatment compartment and the control boxes regulating flow between cells and to the leach drains are visible. The limestone gravel outside is ready to fill up the wetland cells.

How the Wastewater Garden systems looked after the gravel filled the wetland cells. In these early systems, larger rock was used to cover the inlet pipes to prevent odor.

or to the leach drain outside, just like a bath overflowing through the drainage hole when overfilled. We finally figured out how to explain to the crew what we needed. "We are building a swimming pool for the wastewater!" There were murmurs: "*Es cierto! Los gringos están locos!*" The "swimming pool" had to hold the water without any leakage; otherwise we'd have raw sewage going into the groundwater, just like everywhere else in town.

When we fixed all the leaks, it was time to add the limestone gravel. I was initially a bit nervous about this, as some wetland literature had indicated that problems could occur with limestone because it can react with water as well as with some of the compounds present in sewage. I looked for alternatives. Could we get crushed, recycled glass perhaps, or use coconut husks? No. Coconut husks decayed too rapidly in water. Were there other types of rock or gravel available? No. The Yucatan is pure limestone. We would have had to go hundreds of miles to get other types of gravel and that was too expensive. We'd use the local rock. I worried endlessly, during the first few months of operation. It turned out that previous problems

One of the pilot Wastewater Garden systems in Akumal several months after planting. The tops of the walls were tiled and the systems were placed in front of the buildings to emphasize how their beauty enhances the landscape.

with limestone were mostly caused by overloading the system and that these problems seemed to rectify themselves. When I did further research on the effectiveness of our constructed wetlands, I had reason to celebrate that we had in fact used local limestone as the limestone proved great for removing phosphorus.

We were finally ready to plant! Planting is the quickest part of making a wetland. There is no soil in a constructed wetland, just clean, screened gravel 24 to 30 inches (60 to 75 centimeters) deep. We cut the standpipe off at a level that allowed two inches of dry gravel above the water line and filled the wetland with clean tap water. First experience of the strangeness of planting into pure gravel: the pea-sized gravel has a maddening tendency to roll back into the hole you're digging. We put the plants a few inches into the water and kicked gravel back in place around them. We tried a dozen different varieties, including citrus and papaya trees. At the end of this phase we celebrated with a traditional Mexican pig roast, inviting the construction crew and our friends from CEA and Akumal. The experiment was on!

Reka Komaromi, Ingrid Datica and I load plants collected for the new system.

Measuring and Monitoring

We monitored the efficiency of Wastewater Gardens cleaning sewage. That meant laboratory analysis to determine the amount of suspended solids, coliform bacteria and nutrients like nitrogen and phosphorus, in the water. After an unsuccessful attempt to bring samples back to Florida, despite ice-filled containers, we conducted some of the tests which had to be done soon after sampling at labs in Cancun. Other water samples were collected every month by Ingrid Datica and stored until I could take them to the University of Florida.

They were labeled "water samples." I didn't want to have to explain why I had Mexican sewage in my luggage upon arrival at US Customs in Miami, even though I had permits to bring in samples of soil, rock and water. We monitored the water quality in the septic tanks and at each end of the Wastewater Garden compartments. One interesting fact about constructed wetlands is that performance improves after the plants become well-established, unlike a mechanical system such as a car, a computer or TV which, at the get-go, are as functional as they'll ever be. Wetlands on the other hand, like good wines, do improve with age.

Other researchers had set up wetland systems (without plants) and documented that the plants were in fact responsible for some two-thirds of the treatment capability, through direct action and because of root systems that support a wider variety of microbes.

It seemed we were always operating on a shoestring budget. I had studied conventional sewage treatment systems, at the University of Florida and I found it ironic that one such high-tech wonder had been built across from the Center for Wetlands and the Environmental Engineering Sciences buildings, where the systems ecologists had their offices. I was told the university had originally planned to use a constructed wetland for final treatment or "polishing," but had decided to forego it since the sewage plant proved so expensive. The sewage plant, to add insult to injury, had a sludge-drying facility. Several days a week, as a result, it would emanate powerful odors that wafted over to the very building where the wetland ecologists worked!

There was a great lab at the sewage plant and I was offered full access by the director and lab manager. They would analyze the small number of samples I brought in from my two Mexican Wastewater Gardens. Dr. Reddy's lab analyzed my limestone samples and the soils department evaluated my mangrove peat samples. I was in business.

The water quality results were excellent. The data showed we were reducing BOD (biochemical oxygen demand) by 85-90%, nitrogen and phosphorus by around 70%, and fecal coliform bacteria by over 99%, without the use of disinfectants. Fecal coliform bacteria levels were often lower than the levels in the nearby *çenote*. That was auspicious since our wetlands were indeed designed as shit-processing facilities whereas the cenote was simply contaminated by shit floating around in the general Akumal environment.

The amount of phosphorus reduction was surprising too. It was probably due to using limestone gravel. Limestone (calcium carbonate) reacts and binds easily with phosphorus. Even in the smaller of the two wetlands, there was over thirty tons of limestone gravel for microbes to thrive upon and to react with phosphorus.

I begged for and borrowed equipment to perform what sometimes seemed like an interminable number of tests. Light meters were used to measure the amount of sunlight reaching the various strata of our wetland, compared to natural wetlands like the mangrove swamps nearby. We had a fisheye lens camera that could produce images from the ground up so that we could measure the amount of canopy cover developing. A straight stick raised at random was used to measure leaf growth, a simple way to determine increase in the Leaf Area Index. I pumped out the septic tanks, from time to time to measure the amount of wastewater our constructed wetlands received. From the decline in water level I could determine the amount of the wastewater the plants were sucking up.

I did some economic modeling, comparing our system with other sewage systems. I used a unique H.T. Odum-pioneered approach known as emergy which compared the amount of resources used in a Wastewater Garden versus a high-tech sewage treatment system and attempted to assign economic values to both. I counted plants and measured biodiversity. I laid out quadrants of different sizes and randomly measured plant cover. The amount of data grew.

Mangrove Mud

> *Mud, mud glorious mud, there's*
> *nothing quite like it for cooling the blood.*
> –Flanders & Swann, The Hippopotamus Song.

Professors H. T. Odum and Dan Spangler came to visit on a Thanksgiving weekend in 1997, to see the Wastewater Gardens firsthand and advise me on my research program. It was a thrill to observe the legendary H.T. Odum in full swing, scoping out a new ecosystem.

There was torrential rain. A CEA geologist was supposed to take us around but he refused, on account of the weather; H.T. knew we only had a few days to complete our inventory so, into the mangrove swamp we charged! I remember a tribute to H.T. that was given during my first year at Gainesville when he had just announced that he was retiring from academia after five decades. Many former students came to see him, including Bill Mitsch, a professor at Ohio State University and a founder of the Society of Ecological

Akumal, Mexico, 1997. Prof. H.T. Odum demonstrating root depth and peat formation in a mangrove swamp adjacent to the first Wastewater Gardens.

Engineering. A leading wetland scientist himself, Bill showed slide after slide of field trips and asked us what the common denominator was. The answer was H.T. and his students were always up to their knees in water, investigating wetlands. As competent a theorist as H.T. was, he was, first and foremost, a field ecologist. It was exhilarating to be around him because his passionate wish to understand how nature worked was infectious. As we traveled around Akumal together, I could see that H.T. was visualizing systems, seeing how they came together and what the important elements were.

H.T. had worked in scores of mangrove swamps before but he had never seen ones like those in the Yucatan. The limestone here was so porous that there were no surface creeks interchanging water between the mangrove and the ocean. The tide literally came straight up through the limestone and then subsided back down. H.T. could hardly contain his enthusiasm. He also wanted to impress on me that a very good place to send treated water from Wastewater Gardens was into mangrove because it creates peat underneath. So, there was H.T., in the pouring rain, pulling up small mangrove trees to show me their root systems and the peat soil they had produced.

We altered the discharge route of the CEA Wastewater Garden to now go into the mangrove so that could serve as a final biofilter for semi-treated sewage, sucking up the remaining nutrients.

Making that change added a new dimension to my research and dissertation work. Dan Spangler, expert hydrogeologist, advised me how to go about everything from monitoring the magnitude of the water fluxes in the mangrove swamp produced by tides, to the study of the mangrove soil before and after sewage was discharged into it.

On the last day of his visit, Dan drove to the garbage dump which Akumal and the nearby communities used for disposing of their solid waste and other trash. Dan advises mining companies on environmentally sound waste disposal procedures. I had never seen Dan look as shell-shocked and alarmed as when he returned from the dump. The garbage, which included things like batteries, oil drums and containers full of suspect chemicals, had been dumped in an old quarry, a few kilometers from the ocean. He said something like: "I know you're working on the processing of human shit here but what I just saw is catastrophic. Everything that's dumped into that quarry will just leach out serious toxins into the ocean, through the limestone. It's an environmental crime the likes of which I've never seen before." The CEA as well as concerned citizens have been working for years to get the government and the businesses along the coast to do something about this appalling situation, to no avail.

Our new discharge site necessitated that I spend time doing research in the mangroves. The mangroves in Biosphere 2, modeled on the Florida Everglades, had been one of my favorite ecosystems. I used to collect leaf litter and study its decomposition rates.

Mangrove in the Yucatan was dominated by white mangrove but also featured one of my favorite plants, the giant wetland fern (Acrostichum danaefolia) which grows to seven feet. To my utter disbelief, we also found a vine growing in the canopy of the white mangroves that was a member of the cactus genus. It was odd enough to find a cactus that is a vine, but one that thrives in wetlands and is salt tolerant ranks as totally bizarre.

Of course, let us not forget, a mangrove swamp is a hot, humid, mosquito-infested environment, despite the aforementioned charms. I smothered myself with super mosquito repellants that I had mercifully discovered in Australia, and wore Wellington boots and long-sleeved shirts whenever I entered the muddy mangrove realm.

As H.T. Odum had predicted, mangrove soils were ideal for the final treatment of the sewage. When we analyzed the mangrove soils at the labs at the University of Florida, we confirmed they were heavily organic and low in nutrients. This was exactly the type of soil that could act as a sponge to the nutrients. As with so many of the things we tried out, in the course of building our Wastewater Gardens, this was new territory. Before us, there had only been a few experiments using mangrove as an element in sewage treatment.

6
Composting Toilets: A Drier Alternative

"Shit or get off the pot!"

—American colloquial for stop delaying, wasting time and to do it.

AN ALTERNATE TECHNOLOGY to process shit and to prevent pollution is provided by composting toilets. These were being introduced along the Yucatan coast around the same time that we were pioneering Wastewater Gardens.

Dinah Drago, a dynamic Texan, who was also involved in mangrove-protection work, was spearheading an effort to get composting toilets manufactured locally and trying to get the locals to accept the efficacy of the devices. We collaborated with Dinah on a pilot installation at a children's school: Composting toilets were installed there, and a small Wastewater Garden was used for graywater treatment and recycling.

The CEA installed two public composting toilets outside their offices as a demonstration project. There were certain unforeseen problems perhaps unique to doing this in a poor country. These toilets had to be locked at night because otherwise the rolls of toilet paper would be stolen. Excess liquid from these composting toilets was carried away in buckets and taken to our demonstration Wastewater Gardens. Urine is so high in nitrogen that it can "burn" plants if it is applied directly onto plants in a normal garden. In the Wastewater Garden it was diluted with other wastewater and so could be used safely.

Composting toilets in Akumal, Mexico.

When I was doing research on septic tanks for my dissertation, I was fascinated to read about "miracle" microbial additives to "fix" septic tank problems. Many states in the US have been banning their sale since no independent research has ever been done that shows these septic tank additives to be in any way useful or effective. The truth is sewage comes pre-packaged with plenty of microbes that can break it down. People who unthinkingly flush nasty chemicals (paint thinners, solvents, pesticides etc.) down the drain are killing the very necessary microbes that already reside in septic tanks. When a septic tank recovers from such a toxic chemical assault, the microbes recover and resume their work of digesting the sludge. So-called miracle products, offering "super microbes" to rejuvenate your septic tank, appear to work only because the natural microbe population increases once the chemical assault ceases and the input of new shit brings with it a new batch of microbes.

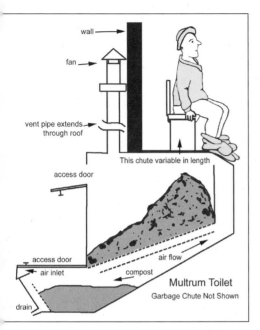

wall

fan

vent pipe extends
through roof

This chute variable in length

access door

access door   air flow

air inlet   compost

Multrum Toilet

Garbage Chute Not Shown

drain

A generic composting toilet (courtesy of Real Goods Company). Some designs use a fan, others just a passive vent, to disperse odors. After each use, shit is covered with organic matter, like straw or peat moss, to further prevent unpleasant smells. Composting toilets can have a composting compartment right under the toilet seat or can send the waste to a separate unit located outdoors or in a basement, for easy removal.

Composting toilets combine the virtues of a traditional outhouse, like those we were using in New Mexico, with a few extra mod cons. They are great in terms of protecting water sources from contamination since they don't mix humanure with water. Some composting toilets are designed to start the composting process right away. One limitation of these systems, however, is that they keep the compost relatively dry, usually by keeping out urine and drawing off any heat by means of vent pipes. That's alright but it means the toilet doesn't generate the high temperatures needed to ensure a complete pathogen kill. There are then two alternatives: keep the compost for a long time, even for a year, to ensure pathogen kill; or move the partially composted material to a standard compost heap, where temperatures can get sufficiently hot to kill any disease-causing organisms. Some designs indeed intend that the waste be removed periodically, so that kind of toilet is really more a collection device than an actual composting toilet. With the installation of small ventilating fans or a well-designed passive vent, the odor associated with old-fashioned outhouses is dispersed, such that a composting toilet can even be installed inside a home. No more night time runs to the "dunny" (Australian slang for a traditional outhouse).

Another drawback with composting toilets is that people have to get involved in the unloading and removal of the shit to complete the composting process. A lot of people aren't ready for that—feces "freak out!" Also, composting toilets can get overloaded with too much liquid from urine, necessitating a safe place to dispose of it. Removal of liquids adds to the time needed to maintain the system. Composting toilets also do not solve problems posed by graywater—the wastewater from the kitchen, laundry, shower or sink requires a recycling or treatment process of its own. Nevertheless, since it is shit that is the most concentrated source of pollution and of pathogens while also being a valuable resource as fertilizer, composting toilets are an important technology. They are available from a number of companies around the world and there are many versions suitable for in-home use. Because the composting process takes place within a sealed container (varying from a simple five gallon bucket to a large system in the basement), potential contamination in areas where groundwater is close to the surface is avoided. The ability to eliminate smells has also made composting toilets more acceptable. There are even designs that come with a micro-flush action, for customers used to hearing the sound of running water after they do their business! Composting toilets mean we deal with our shit at its source.

If we process shit intelligently, in this low-tech yet efficient way, we can ensure, either with sufficient time, or through high temperatures in the composting process, that pathogens are killed. We have also saved the waste and prevented pollution of all the fresh water that would otherwise be needed, and we have avoided all the associated costs in terms of energy and infrastructure. As a final dividend, we recover valuable nutrients in a form safe for direct application back to the soil. Too good to be true? Simple calculations show that even if all seven billion people on the planet wanted and could afford indoor plumbing, using water to flush away their shit, the Earth's supply of freshwater could not meet the demand.

Change is definitely coming. This situation reminds me of a cartoon I saw years ago. A man in his underwear is standing in a flush toilet, holding onto the old-fashioned chain overhead. The caption reads: "Farewell, cruel world!" Well, someday soon it could be the fresh water flush toilet that bids us farewell.

$$\mathcal{C}\!\mathcal{C}\!\mathcal{C}\!\mathcal{C}\ 7$$

Piled Higher and Deeper–
Path to a PhD

"Do bears shit in the woods?"

—An expression indicating the obvious has been said.

WHEN I ARRIVED BACK IN GAINESVILLE, it was not entirely clear what
research project I would undertake to earn my PhD (piled higher and deeper).
Obviously, it would have something to do with waste and wetlands. But Professor
H.T. Odum, my mentor and chairman of my dissertation committee and Dr. Mark Brown,
the co-chairman, were unfazed. These two luminaries were both great believers in self-
organization. Indeed, that's how ecosystems evolve, allowing the mix of competing and
cooperating species to sort it all out, all the while constantly adapting to changing environmental and climatic conditions. If a large asteroid were to hit the Earth causing massive
climate change or a new Ice Age suddenly killed off 80-90% of the living species—not to
worry. Give it some time (in some cases a long time, like tens of thousands or millions of
years) and all kinds of new species will evolve from the remaining few that are left, then
procreate and spread across the planet as before. In fact, this has happened quite a few times
in Earth's history. We have a PhD candidate? Why, surely, his dissertation will self-organize!

I wasn't so sure about this self-organizing process. I wanted to get out of academia
before I was bald and/or senile or bored to death with routine and stasis. Being a graduate

The day of the PhD defense at the University of Florida. Around the table from left: K. Ramesh Reddy, Mark T. Brown, H.T. Odum, me (the defender), Clay Montague and Daniel Spangler. I have made my presentation and all the graduate students in the room are invited to leave before the PhD committee begins its cross-examination.

student had its downside: I was older than most of my professors, for starters, and had already worked for many years on challenging projects with the Institute of Ecotechnics, without the usual red tape, silly pop quizzes, compulsory classes and exams.

As previously mentioned, I had an obligation to build some constructed wetlands down in the Yucatan from where we had collected most of the coral for Biosphere 2's ocean. Akumal, in particular, was in need of affordable, simple sewage processing solutions, especially to protect the coral reef just offshore so off we went to construct wetlands the summer after my first semester at the University of Florida. Later that year, having lunch with H.T. at the faculty club, to discuss the vexing issue of the subject of my PhD thesis, I showed him some photos of the wetlands.

He paused, looked over the table at me and said, "You know you're enrolled in the systems ecology group? Well, looks like you've created a new type of system. Let's study it for your dissertation!" I looked back: "Do you really think we can pull that off?"

Battles in Committee

H.T. Odum and Mark Brown invited other professors to join my dissertation committee. This was a tricky business. Such a committee needs five members of the faculty. It's obviously good to have a diversity of expertise in place, since my research would involve everything from water quality, wetland biogeochemistry, hydrogeology, wetland plants and their ecology, to economics, application, cost effectiveness and computer modeling and simulations.

Academia comes with its own inherent dangers. It is full of strong-minded people convinced that only they know how a particular research project should be undertaken. As soon as the first hint of a storm was brewing, after the initial meeting of my PhD committee, H.T. advised me to keep my research program outline brief (under 10 pages) and not to give too many details. His view was that if you specify precisely what you intend to do, then you already know the answers and aren't open to learning anything new. H.T. wanted to allow for the possibility of pursuing research questions that would inevitably develop later on, giving us the flexibility to follow such opportunities, if and when they arose. This didn't sit well with one member of the committee. In fact, he was livid. I could tell this was so because his face turned beet red and he spluttered with rage. He summoned me to his office and told me he had never seen a dissertation proposal of fewer than fifty pages. He also informed me that he'd spoken up on my behalf since, if my whole research program wasn't outlined in advance, the committee could keep me at it for years simply by demanding additional work. He ended his extraordinary diatribe by wondering aloud as to whether the venerable H.T. had gone senile.

I was shocked, but I knew something was up. I was aware there were serious disagreements in the faculty, punctuated when this committee member came into Mark Brown's classroom one day, ostentatiously put his arm around him and announced that, all rumors notwithstanding, they were the best of friends and had great respect for each other's work. Wow! After a similar scene at another committee meeting, mulling over the direction my research was taking, H.T. simply decided to drop this member from the committee and replace him with a hydrogeologist from the Geology Department, Dr. Dan Spangler. Dan had worked with the Center for Wetlands during the heyday of the Rockefeller Foundation-funded cypress swamp sewage treatment experiments.

My research was broadening and now needed more work on the hydrogeology of the overall region, as well as that of the nearby mangrove swamps. Dan was an ideal addition to the committee. However, dropping someone from a dissertation committee is like lighting a fuse. We prepared for an explosion. Odum, world-renowned scientist and the acknowledged alpha of the wetlands world at the university, advised me to keep well clear of the issue. He said he'd write the necessary letter advising of the change in committee and would handle any fallout himself, since he was the committee chairman.

Diversity versus Monoculture

There are advantages to being inexperienced or an outsider in a particular field. I'd already learned that from running my woodshop back in New Mexico. Unlike veteran carpenters who had apprenticed in the trade, my partner and I didn't know how things "had" to be done, so we were actually freer to try new approaches, experiment with new tools, invent new designs for furniture and make interesting windows with odd-shaped pieces of glass. In short, we could have a lot of fun!

The Biosphere 2 project had similarly attempted and carried out many bold and previously untried things. People working on

(Seated at table from left) K. Ramesh Reddy, Mark T. Brown, and H.T. Odum listen to my dissertation defense. Along the wall are some of my fellow graduate students in Environmental Engineering Sciences and Systems Ecology.

the project were freer to think creatively. There was no book to go by. We invented as we went along. It was the same way when I first started working with constructed wetlands. Only much later, did I find out that most such systems were being installed using just one or two plant species, typically a robust cattail, rush or sedge, which crowded out any competition. Thus it was that many people see subsurface flow constructed wetlands simply as "reedbeds."

Most artificial wetlands are designed and installed by engineers who don't care about diversity, wildlife habitat, aesthetics, beauty or edible produce. They want a simple, reliable, efficient system.

I was naïve enough to think it possible to try and emulate some of the beauty and diversity that I'd been observing in natural wetlands. My simple theory was that a diversity of plants and root systems would improve a system's efficiency. Installing lots of different plants would also lower the chance of a disease or pest damaging the whole system. And anyway, who wants a boring monoculture system in their front yard or in front of the hotel? We could have an attractive system, grow valuable plants, do a good deal of agreeable landscaping, all the while accomplishing the main goal which was the treatment of sewage. In this way, people who'd otherwise be turned off by the industrial austerity of such a system might now *want* to install a constructed wetland.

The Scholarly Tome

I was due to defend my dissertation in late November 1998. Dissertation defense is an academic institution probably dating back to the Middle Ages, as do such pleasantries as the rack and being hung, drawn and quartered. The candidate presents his or her research, which is then followed by questions from any faculty member or student who attends. The public is then asked to leave and, behind closed doors, the committee goes to work on the PhD candidate. It's the committee's last chance to get its licks in.

As I raced upstairs, I searched frantically for a laser pointer. A secretary in the Environmental Engineering Sciences department cheered me up with the comment that dissertation defense was just like a fraternity hazing. But she'd never seen anyone do a defense who didn't eventually get the damned degree.

I didn't find a laser pointer so instead I used an old-fashioned flashlight to highlight my slides. It worked just fine but precipitated gales of laughter from graduate student friends in attendance. After an hour and a half of being grilled on the finer details of my research, I was invited to wait outside, out of earshot in the hall, while the committee deliberated. Eventually Mark Brown came back to tell me to return to the room. He greeted me with "Congratulations, Dr. Nelson." My improbable plan to return to academia to learn more about wetlands was now complete. My work in Arizona and Florida had taken just five years to accomplish since my time at Biosphere 2. I now had—novice tree-planter and compost-maker once known as "Horseshit" or "Landscapes Green"—a doctorate in Environmental Engineering Sciences. I felt as though I had been knighted. I had acquired yet another *persona* (Greek for a *mask*): Dr. Nelson.

Looking apprehensive (and somewhat pale), I await the start of my dissertation defense, November 1998. The scholarly tome lies on the table.

Looking down from a nearby balcony at one of the first Wastewater Gardens in Akumal fifteen years after planting.

8

Doing Business on the Riviera Maya

"Take the shitty end of the stick first."
—Middle Eastern folk saying.

I WENT TO AKUMAL EVERY TWO TO THREE MONTHS, to continue my research, collect samples and monitor the wetlands. In January 1997 we held a seminar at the CEA to showcase the systems. The buildings were not yet fully occupied so the systems were only beginning to get an input of sewage but they were looking good. Since they were perfect for testing which plants would work, I would add a few more plants each time I was there, including bananas and papayas.

Cocktail Parties and the Sniff Test

We also decided to beautify the systems by putting colorful Mexican tiles around the concrete borders. And we held cocktail parties beside the wetlands. We had put Wastewater Gardens in front of the field station and along the main entryway into Akumal—not downwind or miles out of sight, like most sewage treatment facilities. People came and we observed the "sniff test" in action.

The major concern of the citizens of Akumal was what we planned to do about the smell. They couldn't believe that a sewage treatment wetland wouldn't reek of shit. We'd put the wetland less than ten feet away from the main entrance to the PCRF field station. I repeatedly explained to visitors that, in a subsurface flow wetland, so long as the effluent is kept away from air and stays below the surface or under a tight lid (septic tanks and control boxes), there will be no odor. The locals were skeptical. They'd seen the gringos implementing crazy ideas before. But smelling is believing! All day, people would come by to have a look and a sniff. Holding cocktail parties by the wetland was our way of meeting the sniff test challenge. John Allen, the inventor of Biosphere 2, and Deborah Snyder, both fellow directors of the Institute of Ecotechnics, came to visit. John would ostentatiously put a chair near the wetland and spend time each day reading a book, enjoying the fragrance of the tropical flowers and plants in the system. It was Deborah who first came up with the name "Wastewater Gardens."

I recruited Edgar Cabrera, a very knowledgeable Mexican botanist, to complete an inventory of the plants already growing in the two Akumal wetlands. To my surprise, we counted over sixty plant species six months after planting. This was far more than I had originally planted. We figured that additional plants had germinated from seeds in the soils and that birds and other wildlife had brought in seeds as well. This was encouraging, even though we knew species diversity would decline as the system matured.

Some of the plant experiments didn't go well. The citrus tree we had planted with so much optimism, died. Every time I came back to Akumal, I expected to see the remains of the one papaya tree we had planted. It hadn't looked healthy when it went in. In Australia I'd learned just how susceptible papaya is to fungus rot. If the soil was poorly drained or water got onto its trunk, a papaya quickly dies from root or stem rot. But this tree hung on and finally regained its health. The fact that some plants the locals had advised us to grow were not listed as wetland tolerant in any reference book

I consulted, alerted me to the fact that there is still lots to learn. Botanists know what does grow in natural wetlands but they don't know what other plants might have the ability to adapt to and survive in such conditions. I continue to be amused when reading in scholarly tomes, that such and such a plant requires "well-drained soil" when I have seen it doing fine without drainage or aeration in wastewater gardens.

There were other surprises too. Back at the University of Florida there were guffaws when I mentioned that cattails (Typha spp.) were included in my system. "They're overrunning native vegetation in the Everglades because of pollution from cities and farms. Cattails thrive in high nutrient environments. Your systems will be overwhelmed by them." But once taller shrubs and trees began to provide shade, cattail became a much less significant or problematic minor element in our demonstration Wastewater Gardens. Here was another example of ecological self-organization solving a problem.

Plants grew at an amazing rate. Canna lilies grew 8 feet tall. The palms and other trees were growing rapidly too. We'd put in Washingtonia palms. I'd understood the nursery man to say it grows to 25 feet (8 meters). He'd meant 25 meters! In just two years, they were 14 feet (4 meters) tall.

After a year, we had two of the finest-looking gardens in Akumal. Irrigating with brackish water and the effect of the sea breeze makes gardening in Akumal a bit of a challenge but the plants in the Wastewater Gardens flourished. They were getting all the nutrients and water they needed. Akumal's drinking water supply comes from a fairly shallow depth so it is brackish because of the underground mixing of saltwater from the ocean and the freshwater from inland. The salt and contamination by fecal coliform made everyone very "tight-lipped" when taking a shower. Akumal was the first place I'd been where everyone bought bottled water.

The plants grew quite rapidly in the Akumal Wastewater Gardens, taking advantage of the abundant nutrients. The sewage treatment wetlands became some of the loveliest gardens in town.

Explosive Data and Aftershocks

While I was completing my research on the performance of our two pilot Wastewater Gardens, Charles Shaw, the geologist at the CEA, along with some visiting scientists, carried out a simple survey which thoroughly disturbed the good burghers of Akumal. His team went around town testing water for the presence of coliform bacteria. They found surprisingly high levels in the Yal-ku Lagoon, the jewel of Akumal and a major draw for its visitors. This should have been anticipated. There were many houses that fringed Yal-ku and they all had appalling sewage systems. This information elicited a number of reactions from the locals. I was eerily reminded of the doctor in Ibsen's play, *An Enemy of the People,* who is vilified when he finds evidence that the town's therapeutic hot springs, whose visitors support its economy, are seriously polluted by wastes from a tannery. So it was in Akumal, where many preferred to see no evil, certainly smell no evil, and never, ever even speak about evil (shit) in their backyard!

There was an alternative reaction of genuine resolve amongst others that it was time to do something, especially before the truth got out. Suddenly, home-owners and hoteliers began making their way to our office to inquire about the viability of these "Wastewater Gardens" they'd heard about. Business took off.

PCRF started a Wastewater Garden Division, which I headed. We'd started our wastewater cleansing business in response to demand in May 1998, less than two years after the first Wastewater Gardens were constructed. Laura Bush, a scion of one of Akumal's oldest families and the owner of the Villas Akumal Hotel, right off the main beach, decided to install a system to treat sewage from her outlying bungalows. Fortunately, the site was sandy so we were easily able to place the system deep enough for gravity-flow of feeder

Part of the PCRF team in Mexico: from left to right, John Allen, Abigail Alling, Deborah Snyder, Gerard Houghton, Mark Van Thillo and Ingrid Datica.

pipes from three septic tanks that fed into one large Wastewater Garden. The hotel replicated the diversity of plants we had installed in our prototypes. The system flourished and to my amazement, a year later, they even had papayas hanging off their happy, healthy trees! Excellent bananas were produced too.

We invited our friends, Klaus Eiberle and Reka Komaromi, to assist us in our new business. Gonzalo Arcila was appointed director of operations, Klaus was the construction supervisor, Mark Van Thillo an invaluable troubleshooter, and Reka and Ingrid Datica took care of the planting. PCRF trademarked "Wastewater Gardens" and we had people sign confidentiality agreements in order to protect the intellectual capital of our innovations in constructed wetland technology.

The news of coliform pollution in the Yal-ku Lagoon brought increased attention to the concept of constructed wetlands. We had installed systems at a few homes, after the Villas Akumal Hotel system was put in. Some clients were actually motivated simply to have a nice looking garden. The systems could also serve other functions: one house used the Wastewater Garden as a privacy, noise and dust

An L-shaped Wastewater Garden for a house in Akumal, Mexico.

Wastewater Garden for the kitchen at Xpu-Ha, where space constraints dictated a triangular-shaped system.

Raised Wastewater Garden for home adjoining Yal-ku Lagoon in Akumal, Mexico

barrier from road traffic. For these purposes, we built a long narrow system on the edge of their property. Dominated by the wetland fern, Acrostichum and canna lilies, it quickly formed an eight-foot high green barrier.

We designed systems that fit unusual property layouts. We built L-shaped systems to wrap around the outside of buildings or walls; kidney-shaped ones, just for the sculptural aesthetic, and systems that filled up whatever green space was available, from a sidewalk right up to the front door of a home or condominium.

Eventually, we decided to offer Akumal homeowners Wastewater Gardens almost at cost, as a civic gesture. We barely made a couple of hundred dollars profit per installed system. But, despite the euphoria and sudden flowering of interest there were some growing pains and simple reminders about the reality of doing business in Mexico. Things do not work according to the stereotypical Anglo-Saxon, on time, workaholic mentality. This is the land of *mañana* (tomorrow). *Mañana* – all will happen someday, certainly not now, but in the fullness of time. A few more mañanas were needed for delivery of materials, for each step of the construction or sometimes even for the reappearance of the construction crew who might have gone back to another of their unfinished projects. We had plenty of time to work on the virtue of patience.

We installed Wastewater Gardens at most of the private houses bordering the Yal-ku Lagoon. Residents were elated when a new set of measurements showed coliform contamination levels were now much lower.

The largest integrated Wastewater Garden system we installed was for a new Eco-park, Xpu-Ha, 12 miles (20 kilometers) north of Akumal. Our system was designed for up to 1500 visitors per day. This ecological theme park included animal displays, snorkel-

ing and diving, restaurants and bars off the beach and beautiful pathways, on a site of around fifty acres (20 hectares). Engineers told us that the cost of the Wastewater Gardens was less than what they would have spent just collecting and pumping wastewater to a conventional treatment plant. We built constructed wetlands for the entrance area, the kitchen, the veterinary clinic, and several for the beach restaurants. I was especially fond of the gardens by the restaurants. Customers could be drinking their margaritas and, in theory, realize the impact of their trips to the toilet, manifested by the beautiful plants growing before their eyes, nearby.

A Wastewater Garden provides landscaping around one of the palapa-roofed restaurants at Xpu-Ha Ecopark, Mexico.

Imitation is the Sincerest Form of Flattery

Before too long, of course, other people tried to build systems like ours but many lacked the requisite experience and expertise. I realized once more how apparently simple yet complex a Wastewater Garden is, when I learned of the myriad ways people who don't know what they are doing cut corners, do things on the cheap and generally screw up.

The first misunderstanding was that the plants are not just there for decoration. They are a vital part of the treatment process. We watched as people selected their own plants, without any understanding of their usefulness. I was aghast to see systems that contained only the daintiest and least robust plants on our species list. Despite our fervent explanations, folks didn't appear to understand that plant root systems are the aerating pumps of the treatment process. Nor did they understand that the right plants are needed to take up nutrients from the sewage and that they support a variety of essential microbes in their roots. I attempted to correct this situation by advising our team to maintain control of plant selection and offer a "Chinese restaurant menu" of "shallow," "medium" and "deep-rooted" plants, a combination of which needed to be used: two from "Column A," three from "Column B," four from "Column C," and so on.

In addition, there were largely ill-fated attempts to simply flat out copy our design. One hotel director picked our brains about our systems; then went ahead and had a wetland compartment structure built but never completed it. Apparently, the necessity of drops in elevation to enable wastewater to flow downwards, without the use of pumps, was beyond the competence of their architect. It certainly wasn't obvious to casual imitators.

The most egregious examples were where imitators would build a system that sort of resembled a Wastewater Garden but was clearly too small to handle the amount of wastewater coming from the project's building. Just having a constructed wetland doesn't guarantee that it has been sized properly and will do what it should. If one wasn't there for the leak test, there's also no guarantee that the system in fact really holds the wastewater before it is discharged. Think of the septic tanks that are unlined but which to outward appearances look like the real thing.

In spite of all this, I wasn't too concerned or upset. It's part of the landscape when you're working in the real world. I figured since we were the first to bring subsurface constructed wetlands to the Yucatan, that it was inevitable and even desirable that other companies would later try to emulate us. After all, there's plenty of shit to be treated—certainly enough for all the startup businesses and potential competitors.

The Riviera Maya and the Building Boom

Of far greater concern to us than our imitators was the dizzying pace of residential and commercial development along the Yucatan coast. Mexican authorities had rechristened the coast from Cancun south to Tulum and the Biosphere Reserve at Sian Ka'an, the "Riviera Maya." The dollar signs were like neon lights in everyone's eyes. There was a workshop at the CEA where it was noted that the town of Playa del Carmen was now the fastest growing urban area in all of North America. Playa has been transformed from a backwater fishing village and hippie beach town to a sprawling town with more than 194,000 inhabitants—and is still growing. This rapid growth was entirely without any coherent town planning, infrastructure development or, no surprise here, any sewage treatment plan. There was a tiny sewage treatment system that the phantom septic tank trucks were supposed to find, but very little of Playa del Carmen was connected to it. The sprawl was full-throttle on all sides, radiating from

The number of hotel rooms along the 80 mile coastline of the Rivera Maya went from 12,600 in 1999 to 40,800 in 2014 to accommodate the 3.8 million visitors each year to the region.

the original core beach areas, and even spreading through former mangrove and forest areas. It was all being filled in, bulldozed down.

The main highway from Cancun to Playa del Carmen was widened to two lanes in both directions, and there was talk of creating new towns of 10,000 to 30,000 inhabitants each, to house all the workers needed to service the new tourist meccas.

Clearly, as a result, mangrove swamps along the Yucatan coast are under assault. "Protected by law" in Mexico doesn't mean that much. Every time I came back to Akumal, it seemed a hotel or condominium owner had decided to create more parking spaces and so brought in a load of soil or gravel to cover mangrove areas which fronted their street. I had been told destroying mangroves carries a heavy fine under Mexican environmental regulations, but a *mordida* (bribe) usually takes care of that. Elsewhere along the coast, large resort developments simply drained the mangrove swamps and infilled.

The mangroves are a vital biofilter protecting the marine environment from the flow of nutrients carried in Yucatan's underground water systems. Charles Shaw, the geologist at the CEA, shared the results of his hydrogeology research. In river estuaries, freshwater from inland mixes with saltwater from the ocean, on the surface. In the Yucatan, this process takes place just below the surface, underground, because of the porous limestone geology. Mangroves, which generally grow just inland from the beach and sand dunes, are the equivalent of coastal wetlands. They purify the water from the land before it gets to the sea. With the loss of the mangrove line of defense, the impact of nutrient pollution on the coral reefs is even more severe. Thus I was delighted that our research eloquently demonstrated the role mangroves can play in a sewage treatment system.

So much construction and development was going on in Akumal that the workers preferred simply to go into the mangroves

Viva Mexico!

Villa Balamek, Private Residence, Akumal

Our first system
built in 1996
as it looks today!

Maya Tulum

Maya Tulum, Yoga Retreat Center, Tulum

Xpu-Ha EcoPark, Riviera Maya ... 1500
visitors a day

GREENLAND
(Danish)
ICELAND (Danish)
NORWAY
UNITED
KINGDOM
FRANCE
SPAIN
PORTUGAL
Gibraltar
(Br.)
ALGERIA
MOROCCO
RIO DE ORO
TUNISIA
AFRICA
CANADA
Newfoundland
UNITED
STATES
Azores
(Port.)
Madeira
(Port.)
Canary Is.
(Sp.)
Bermuda
(Br.)
Gulf of
Mexico
Bahama Is.
(Br.)
MEXICO
CUBA
PUERTO RICO
WEST INDIES
Caribbean Sea
Cape Verde Is.
(Port.)
GAMBIA
PORT. GUINEA
SIERRA LEONE
LIBERIA
CENTRAL
AMERICA
PANAMA
CANAL
VENEZUELA
GUIANAS
COLOMBIA
Galapagos Is.
ECUADOR
Atlantic
Ocean
BRAZIL
Ascension
(Br.)
PERU
St. Helena
(Br.)
BOLIVIA
Pacific
Ocean
PARAGUAY
CHILE
URUGUAY
ARGENTINA
Falkland Is.
(Br.)
Graham Land

Wastewater

MAP LOCATIONS

MEXICO: Yucatan coast from Cancun to Tulum
ALGERIA: Temacine
WESTERN AUSTRALIA: Derby, Fitzroy Crossing, Kununurra, Broome
BELIZE: Stann Creek, Dangriga
SPAIN: Malaga
PORTUGAL: Tavira, Algarve
INDONESIA: Aceh, Java, Bali, Sulawesi, Bunaken Marine Reserve
THE PHILIPPINES: Manila
FRANCE: Aix-en-Provence
PUERTO RICO: Patillas
THE BAHAMAS: Cape Eleuthera Island
POLAND: Lutowiska, Tri-Lateral Biosphere Reserve, Magursky National Park, Krakow
IRAQ: (planned) Al-Chibaish, Marsh Arab towns around Mesopotamian National Park

Garden Sites

Where It All Began ...
Synergia Ranch–Santa Fe, New Mexico

Organic garden and orchard at Synergia Ranch

WWG at the Ranch

Biosphere 2

Biosphere 2, Oracle, Arizona

Section of the wastewater recycling system

Our half-acre farm

The Green Jewel of Bali

Tirtagangga Royal Water Gardens

Two Wastewater Garden Systems
handle the effluent from the visitors

Sacred Mountain Sanctuary, Sidemen

Sunrise School, Kuta/Legian

Private Residence, Batuan

Villa Complex

Adventures Down Under

Birdwood Downs, Derby, WA

Freda Wilson's house system,
Emu Creek, WA

Ned Johns's house system,
Emu Creek, West Australia

Joy Springs, Fitzroy Crossing

Coco Eco B & B, Broome

Island Living

Puerto Rico

Las Casas de la Selva, Patillas

The Bahamas

Cape Eleuthera

Belize

Kanantik Reef and Jungle Resort

Oasis in Algeria

Crescent Moon system, Temacine

Iraq

Next stop, Iraq!

Conceptual design for Eden in Iraq

to relieve themselves rather than take a long walk to a bathroom or composting toilet. I found it ironic that the Maya were intuitively heeding their calls of nature in the most ecologically responsible manner. Mangrove soil is the best place for their shit. (Although I did have to step carefully when I went into the mangroves to conduct research on how they were working as the systems' final purifier!) Using the mangroves was being far more responsible than the grandee expatriates, who sent their business to the unlined septic tanks behind their luxurious houses and thence into the ocean.

In this helter-skelter of development, little heed was being paid to the environmental consequences. An extremely large development slated for Puerto Morelos would have destroyed large areas of mangrove. Environmental groups protested to the Quintana Roo state government. It was later discovered that the governor himself was a silent partner in the project. An almost happy ending to that story is that he was later indicted on cocaine-trafficking charges and went into hiding.

Mexico's admirably strict environmental statutes are often vitiated by the corruption of some of its officials. There are also those uniquely Mexican takes on environmental regulation, such as the stipulation that a hotel's fresh water and sewage only be tested *before* a treatment plant is installed. If a sewage treatment system is already in place a hotel is exempt from testing! As a result there is a "Potemkin village" syndrome, where an appearance of effective sewage treatment is maintained while equipment is neither maintained nor correctly operated. Most big developments in the Yucatan are financed from afar or by multinational corporations based overseas. And it's overseas where the profits are sent.

There are some who say that in developing countries, the economy must be grown at all costs, and that environmental regulation is a luxury. This is nonsense, of course, since it is the poorer members of society who suffer most from environmental degradation. Any so-called "development" that destroys natural resources and damages the environment will eventually also destroy the well-being of its human inhabitants. The wealthy can afford to go to unpolluted spots or shield themselves from ugliness and unpleasantness.

Our small scale business also ran into difficulties obtaining permits for a project in the south of the state. Mystified, Gonzalo tried to discover what was going on. The drillers who put in deep injection wells apparently saw us as a threat to their lucrative if useless business and so they had approached the people who issue permits. We had to go to the state capital in Chetumal to make presentations to the government and the local university, before things were rectified.

Raising Awareness

I was really curious to see whether interacting with a Wastewater Garden was having the desired effect of making users more aware of their choices and actions. There are so many things we take for granted about the water we use and get rid of, and there is also a great deal of ignorance. For instance, when we were putting in the first demonstration systems, people were concerned that the plants in the systems would be damaged by bleach. Bleach is used in greater quantities than normal in Akumal in order to remove the salt residues deposited by ocean breezes. Bleach is dilute hydrochloric acid. It is thought of as dangerous but this same acid is present in our stomachs and helps us break down and digest food. Wastewater is a mixture of water from the toilet, the shower and the kitchen sinks. Any bleach used will be considerably diluted. So, in fact, this worry about bleach was irrelevant.

Mostly, people did get it. For example, there was an Italian chef, whose restaurant and beach hotel, south of Tulum, were using our system. He called us because he was worried about using some

An Italian restaurant and resort along the coast in Tulum, Mexico. The Wastewater Garden is the first thing that visitors see when they arrive.

new oven cleaners and toilet fresheners. He wanted to know if they would harm the plants which he loved in his Wastewater Garden. He was so keen on the systems that he experimented and found some of the herbal plants he used for cooking could grow there.

People were also concerned about toilet paper. Will it harm a Wastewater Garden system? Septic tanks can deal with normal inputs of toilet paper because it actually breaks down quickly. In Biosphere 2 we had managed without toilet paper altogether. We found special fittings used in Middle Eastern countries, a little spray attachment at the end of a flexible hose, to give our asses a quick cleaning spritz of water. But that decision was mostly based on the goal of Biosphere 2 to reduce or eliminate "consumables," products which couldn't be produced inside the facility and would have to be imported from outside. There are consumer products that indeed are far more inappropriate in terms of disposal into a septic tank or Wastewater Garden or other constructed wetland sewage treatment system. Just because it is a liquid doesn't mean that it is okay to pour it down a drain.

In Retrospect

Between 1998 and 2000 we built more than fifty Wastewater Gardens in the Yucatan. Our most significant efforts were in Akumal where we had started. By 2000, almost sixty percent of the town's residences and businesses were connected to Wastewater Gardens or constructed wetlands built by our imitators and competitors.

Installing constructed wetlands for sewage treatment is a new and alien concept for most people, despite a long track record of success around the world, spanning decades. Perhaps, as I have suggested before, clients would be reassured if there were high-tech monitoring systems, with flashing lights and controls. Modern society believes in technology. The more "advanced" it is, presumably the better. Wastewater Gardens are perhaps seen as too natural or too simple to be really effective. This perception then leads to the erroneous conclusion that there is no *real* engineering or science involved.

Another interesting phenomenon one encounters in this business is that many people (especially wealthy gringos) resent having to pay for a system at all. Maybe they think since we are dealing with environmental problems that somehow the government or some well-financed, do-gooding, idealistic eco-group is going to pick up the bill for cleaning up their shit.

I recall one homeowner, a retired doctor, who was upset that we were making a profit building Wastewater Gardens in Akumal. He said that since it was a question of environmental protection, we should be offering a free service. It only occurred to me later that one could respond that since medical treatment is a health issue, it should be free too. Had that been the case, our retiree would not have been able to afford his lovely home in Akumal. As far as I know, this gentleman has stuck to his guns and has still not installed any kind of suitable sewage treatment system at his house or rental guest house property. His shit is thus still percolating through the limestone, virtually untreated, into the groundwater and on out to the coral reef. May he enjoy many a lengthy swim in his own company!

Sometimes Silence is the Best Compliment

In time, the original area where we built our first two trial wastewater gardens, became beautifully landscaped and a number of new buildings were put up and new enterprises started. On one visit to Akumal, I sat in the outdoor patio of a new bakery which overlooks one of the Wastewater Gardens. Although there were small signs in Spanish and English that here was a "green" sewage treatment system, people at the tables hadn't seen them. Sipping their cappuccinos and eating their croissants, I overheard comments about the lovely tropical garden, with its tall palm trees and abundance of flowering shrubs, offering shade. I thought of those joggers going around the park in Orlando, Florida, where shit was being transformed into flowers and greenery, quietly and with pleasant odors. Perhaps the best compliment in this business is when no one realizes what's happening right under their noses!

Every morning I went out to do battle with the invasive wattle trees, making
sure to dig out all the side roots.

9

Outback Yarns and the Nitty-Gritty of Building Wastewater Gardens

"The natural history of Australia is so strange,
there's no need for fiction there."
—Mark Twain.

Can You Dig It?

I was going deeper and deeper, but it was getting more difficult now to shovel soil out of the hole. I pulled out my tape measure. It wasn't deep enough yet, so I kept digging.

It was April Down Under, the start of the cool, dry season in 2000. I was digging our first Wastewater Garden in Australia, just behind the homestead at Birdwood Downs, in the remote and sparsely populated Kimberley region of the North West. I was back at the tropical savannah project I had helped start in the late 1970s. It is one of a series of "demonstration projects" consulted by the Institute of Ecotechnics, of which I have served as chairman since 1982. These biomic projects are deliberately set in regions of the world where challenging conditions mean that conventional approaches do not work.

The Kimberley region was settled by Europeans just over a century ago. Vast areas were overgrazed by sheep and cattle. When such tropical land is degraded, it gets invaded by dense thickets of "wattle" (short-lived Acacia trees). In places farther inland, the land had turned into near desert. At Birdwood Downs we had to deal with 15,000 wattles that

Improved pastures at Birdwood Downs, with Boab, Eucalyptus and other native trees providing shade for animals. Cleared of invasive weed trees like Acacia (wattle), the beauty of this ancient landscape is revealed.

could regrow on an acre of land, in a climate that alternates between wet (for about four months) and dry, where typically there is no rain for seven to eight months.

It's a land of extremes and catastrophes, from locust swarms, to floods, to bushfires, which can rage across a thirty-mile front. In the hot season, you can die of dehydration in twenty-four hours, which is another good reason why the Kimberley, which is larger than Germany, has a permanent population of around 40,000!

By 1984, we'd managed to plant improved drought-resistant grass and legume pasture on half of Birdwood Downs (the Australian government's requirement to gain freehold title over the 5,000-acre [2,000-hectare] property). The land was sold to us at a rock-bottom price, as the government was eager for us to succeed because there were tens of millions more of similarly overgrazed and almost worthless acres of land in the Kimberley.

We hadn't done much with manure in those early days at Birdwood Downs, just some composting as we started to create a garden oasis around our homestead. We did put in a couple of septic tanks and let one drain into and feed our banana patch. Perhaps it was a sign of things to come that I was greeted as "seppo" in Kimberley pubs. A little sleuthing of "Strine" (Australian slang) revealed the etymology of this phrase imported from cockney rhyming slang. A Yank (an American) is a "septic tank," but since Aussies eschew excessive syllables (refugee = reffo; journalist = journo) this is shortened to "seppo." It was time to bring what I'd learned about wastewater treatment to a land and region I love.

Among the numerous challenges we faced was a soil severely depleted from the decades of overgrazing. There was no topsoil to speak of when we started. The good news was that this soil, compared to the concrete-like clay soils of New Mexico, was easy to dig, especially when it was still moist from recent summer wet season rain.

The Birdwood Downs homestead building has a two-chamber concrete septic tank, which I had helped build twenty years earlier. The septic tank was deep enough to receive wastewater from the kitchen, toilet, and shower. But to have a pump-free system, we needed to make the Wastewater Garden a "sunken garden" so that sewage would run into it by gravity flow from the septic tank. That added an extra foot and a half to the normal 3 feet (0.9 meter) depth needed. Since the system was just 7 feet wide by 13 feet long (2 by 4 meters) and the soil easy to dig, it all seemed simple. I was delighted we had figured out how to avoid using a pump. I grabbed a shovel and began digging with enthusiasm.

After that initial euphoria subsided, I had to simply become determined for, as one might expect, as I went deeper and deeper, the soil got harder and harder. At first I could shovel soil right into the trailer I was using to haul it away. Now I had to work even harder to get what I dug back to the surface. A second shoveling was now required to get the soil actually into the trailer.

I consoled myself with two thoughts. I loved being at Birdwood Downs because I got to do physical work like wattle chopping or

Using an old fashioned shovel, I am excavating our first Australian Wastewater Garden in back of the Birdwood Downs homestead, Kimberley region of Western Australia.

In the early years at Birdwood Downs, we operated as a seed company, since we needed seed for regenerating the land and the income helped pay for development. I am filling up bags of Birdwood grass seed after our harvest went through a seed cleaner.

haying, and so burned off layers of fat my midriff had accumulated from spending too much time being sedentary, reading, or in front of a computer. Secondly, the digging reminded me of an ecotechnic exercise back in the early Synergia Ranch days to first do things the old-fashioned way before employing a machine. For example, we'd cut all the wood for a while with a handsaw before using a power saw, which made us appreciate what the machine was doing. So here I was hand-digging, over the course of a couple of weeks, a hole that a backhoe could excavate in a mere hour or two. But I was too far into it to stop and call in the machine. I was also proud—and a lot slimmer.

Pilots and Legal Limbo

Australia is a huge country with a small population, a large percentage of which seems to be employed by one government bureaucracy or another. Western Australia is known as "W.A." Some people, especially those trying to get something done that requires a permit, claim that W.A. stands for "Wait Awhile." This is the Australian version of *mañana*. It will get done, but not right now—maybe after the next long weekend or after the "wet" or the "dry" or after we get back from our holidays. But by then a lot of other work will have piled up on the desk and there'll be a whole new set of regulations that no one really understands but with which we will need to comply.

Australia is also a very dry country. Only Antarctica is dryer even though it's covered in ice. Australia also has its fair share of sewage treatment and disposal problems, especially in remote areas and in the aboriginal communities that make up a large portion of the population in the "bush" or "outback." One would assume that innovative approaches to water conservation and recycling would be welcome. Well…yes and no.

We introduced our work with constructed wetlands at meetings with numerous state officials, from the local shire (county), which in our case is the size of a large European country, to state officials in Perth. They were quite positive and supportive.

The need for good sewage treatment is acute in the outback. High-tech sewage treatment requires a lot of complex machinery, which means trained personnel and easy access to spare parts and chemicals. It also costs lots of money. This approach is also un-economic if populations are small or widely dispersed. The larg-est town in our region, at that time, was Broome, with fewer than 5,000 people. Much of the population is on septic tank systems. This is a problem during the wet season, as so much of the soil gets water-logged, causing leach drains to stop working, which in turn allows untreated wastewater to surface. Our nearby town of Derby, for example, can smell strongly of sewage at that time of year. Other places had problems because of high water tables or tidal flow. The coast off the Kimberley has the second highest tidal range in the world—thirty-five feet (11m.) between low tide and high tide. This means coastal communities are subject to extreme fluctuations in the depth of their underground water. Add to that a propensity of many people, white or aboriginal, to situate their houses near natu-ral wetlands, and you run into difficulties getting a septic tank and leach drain system to work year-round.

The sewage treatment option for aboriginal communities and outback towns that is currently favored by the W.A. Health De-partment is sewage lagoons—large ponds where natural mecha-nisms treat the sewage. The biggest problem with lagoons is that they need an extensive and expensive pipe and pumping system to get the wastewater from the houses to the lagoons. Operating these systems also requires trained personnel and a reliable and uninter-rupted supply of electricity. Power lines often come down during the wet season's strong winds, and lightning strikes can cut off electrical supply. We drive by sewage lagoons in Mownajum, a nearby aboriginal community, on our 20-kilometer commute into Derby. The smell can be very noticeable, even in a fast-moving car.

Setting up sewage lagoons is expensive. One community of around 300 people had a sewage lagoon system that cost $1.2 million Australian dollars to build—that's $4,000 per inhabitant. Many outback communities number just a few dozen people. Sewage lagoons add little to the aesthetics of a community, and unless you're a fan of green algae, are not very interesting ecologically. They're part of the "get wastewater away from people and treat the sewage" syndrome, not a "treat, recycle, and reuse" paradigm. There's not one example of treated water from the lagoons being used for irrigation. And, once installed, despite high fences and locked gates, children from the community often climb in and use sewage lagoons as swimming holes.

So we received approval from the Department of Health, despite their attitude that they would ignore all the history of constructed wetlands even in other parts of Australia, to install a series of Wastewater Garden pilot projects. The first two were planned for the area around the jetty at Derby and at the homestead at Birdwood Downs.

The jetty is about 5/8 mile (one kilometer) from the center of town and a favorite spot for locals to fish and catch crab. It's a nice place to watch the sun set over the water at high tide—or at low tide to look at the peculiar moon-like terrain of King Sound. There was also a great seafood restaurant there. The Wastewater Garden was designed to serve a new public toilet and the restaurant's toilets. It would mean there would be a beautiful tropical garden at one end of Derby, and it would protect the mangrove and coastal waters from sewage pollution. I say *would* because while I was away,

I received some puzzling e-mails telling me our project had been abandoned for a most peculiar reason. The Shire engineer, and engineers 1,500 miles (2,400 kilometers) away to the south in Perth, the state capital, were afraid that the septic tank needed for the project would sink into the saltwater below the location of the excavation. My response—that freshwater (like sewage) is lighter than saltwater, and the septic tank could be anchored more firmly—was ignored. The real issue probably was that none of the engineers had ever seen a constructed wetland. It was too exotic for a civil servant to approve. So, the Shire engineer went ahead and had a kilometer of pipe laid to connect the jetty and to pump sewage into Derby's already over-loaded lagoon system at the other end of town. He felt chastised later when the National Heritage Trust later approved our funding proposal and offered to contribute some 50% of the cost of the system. But it was too late; his pipe had been laid.

We went ahead with the other pilot project and constructed the Wastewater Garden at Birdwood Downs. We also did two years of water quality tests, which confirmed that, as expected, the systems were working extremely well. We reduced organic compounds (BOD), resulting in 50% more cleanliness than the levels mandated for municipal systems. We had a 98% reduction in fecal coliform bacteria without using a disinfectant and also had good levels of nitrogen and phosphorus uptake. The garden is also quite beautiful. We have giant taro, purple elephant ear, three colors of canna lilies, heliconia (bird of paradise), flowering oleander, papyrus, ginger, cardamom, coconut, and jasmine growing in the system. All that greenery was inviting. One day we had to rescue a horse that was grazing by one of the Wastewater Gardens and slipped down the eighteen-inch drop into the system. Neither horse nor wetland was injured. We installed a paperbark post and rail fence to keep our curious and hungry four-legged friends at a distance.

The Birdwood Downs Waste-water Garden after a few years growth. This garden, lush with canna lilies, bird of paradise flowers, ginger, coconut palm and oleander, is the system I was threatened with imprisonment for creating.

I was more bemused than alarmed when, at a meeting at department offices, more than three years after the Birdwood Downs system started operating, I was told that I did not actually have final approval for the pilot project. I replied that I had answered all the Health Department questions, submitted plans to the Shire and had its environmental health officer inspect the system as it was being installed.

We had done two years of testing, as per pilot project requirements, and I was so pleased with performance that I invited everyone I could think of to come and inspect the system. Oh No! I didn't have a certain piece of paper. I was told that this was a serious matter. Without it, I could be fined or even go to prison. What was amusing was that by now I had been to lots of sewage treatment systems around the Kimberley that were fully approved "standard systems," but which were, in my opinion, unmitigated disasters and a public health menace the moment they were installed. I contemplated being sent to prison for the crime of turning shit into flowers and clean water!

Eventually everyone cooled down (though I relished the free publicity I might have gotten if handcuffed and taken to jail), and I had further friendlier chats with the Department of Health people. We needed approval for other systems. I suggested that the setup at Birdwood Downs be approved since, one, I hadn't known I needed a specific form and two, the system had been already working flawlessly for nearly four years. "Oh no," I was told, "we can only approve something *before* it's done. We can't give approval for a system after the fact. However, since it's working fine, we'll just ignore that."

Applications on Aboriginal Communities: Mud Maps and Dreamtime Control Boxes

Discussions about our first aboriginal community Wastewater Garden project had spawned piles of submissions, a plethora of e-mails, photos of low-lying houses during the last big *wet*, worst case scenarios and assessments of the depth of potential flooding around the houses. We were at Gulgagulganeng (Emu Creek) near Kununurra, in the East Kimberley. We sat in the dirt making a "mud map" of the community, pointing out where new Wastewater Gardens might be placed. Everyone was there: the local plumber, the shire Environmental Health Officer, the project engineer from Perth, administrators from the local aboriginal resource agency, and community members. The mud map was a far more effective way to communicate than a PowerPoint presentation or books of photographs, maps, or even verbal explanations. Objects close to hand were used to represent things. This being a somewhat alcoholically dysfunctional community, this meant empty beer cans and plastic wine flagons and of course the ubiquitous cigarette packets. But the mud map finally got everyone's attention and illustrated the plan very well. "What's the Winfield packet?" "That's Ned's house?" "Oh, yeah, near the wine flagon, next to Freda's ..."

Community meeting at Emu Creek where Wastewater Gardens were approved to solve sewage problems. Kneeling is Robyn Tredwell of Birdwood Downs, to her left, Kate West of Arup. Everyone pitched in to make a mud map of the community, with objects at hand, to make clear where the Wastewater Garden systems would be located.

The community had turned out even though it was a national holiday, ANZAC Day. Everyone was sober. This was serious business. For years, Emu Creek had suffered from poor health because septic tanks were installed too low in the ground. Half the homes were near natural wetlands or in areas that collected standing water when the nearby Emu Creek flooded. Local plumbers had been making a fortune having to pump out the septic tanks every week or so. Sometimes the ground was so soggy even in the dry season that the septic tanks would get backed up with groundwater. So, the septic tanks were sealed and separated from the leach drains and pumped out when full. In short, Emu Creek was a great example of the poor community planning, bad design, and weak infrastructure that is sadly too common in aboriginal communities up north. The community members were moved here six miles out of town because their original neighborhood closer to town was seen as an eyesore. So it was decided to move the people and their drinking problems away. These were the "throw-away people," unwanted "black fellers" that a town is only too happy to relocate to the bush, so as to avoid offending the sensibilities of tourists who come for a few months during the cool dry season.

Yet they were wonderful people, except when drinking. Some Gulgagulganeng residents are outstanding aboriginal artists and internationally known. The natural setting of the community is beautiful, with magnificent trees surrounding the houses. I didn't notice at first that these giant trees were paperbarks (Melaleuca spp.) because they were far taller than the ones we had back in Derby. Paperbarks grow naturally in wetlands. If anyone had "read the country" or noticed the plant species growing here, they would have known how wet the ground can get. It was madness to put houses here, and to make sure the plumbing didn't work, the septic tanks were placed unnecessarily deep in the ground. This left one of two conclusions: Either the original plumbers were incompetent, or deliberately planning to make a steady income from dealing with them since septic tanks that deep couldn't possibly work properly. Clearly, some of the state and federal money destined for aboriginal communities would have to be diverted to pay for repeated septic tank emptying.

Such was the situation when we received an e-mail from Kate West, a smart young engineer from the Perth office of ARUP, an international engineering firm. She was working on the Emu Creek community project and had been told not only to fix the situation but, if possible, find sustainable solutions.

Somehow, a paper written about Wastewater Gardens, authored by me and Robyn Tredwell (who managed Birdwood Downs for twenty-five years until her untimely death in 2012), had landed on her desk.

So, Robyn and I went to the community to conduct a feasibility study on how we might fix the sewage mess. From Derby to Kununurra is a mere 600-mile (960-kilometer) drive. It was mid-December, early in the wet season, and there had recently been 4 inches (10 centimeters) of rain. This was a stroke of luck as it

Robyn Tredwell, General Manager of Birdwood Downs, standing in front of the Wastewater Garden which treats sewage from the homestead building.

became clear that the "Central Park" at Emu Creek, on the inside of the ring road connecting the houses, was on higher ground and had coarse sandy soil that showed no evidence of ever having been waterlogged. Even after the rain, the water-holding capacity of the soil in the Central Park was excellent. We could send wastewater from the lowest lying houses out to the Central Park without relying on pumps year-round. Here would be a variant of the "sunken gardens" at Birdwood Downs. It would be a challenge. We would probably have to dig deep and build high berms to keep wet season floodwaters out.

When the community understood its sewage would now be treated and that in place of raw sewage on the ground and kids getting sick, beautiful gardens would grow, they were delighted. What kind of plants did they want? The "white feller" government workers always talked about using native plants, but the answer from the community was loud and clear: "pretty flowers and fruit trees that we can harvest and eat from." Once the Wastewater Gardens were installed there would be plenty of bulbs and shoots that could be planted around the houses to continue the greening of the community. Greening aboriginal communities is important for health as well as improving the quality of life. There is a high incidence of eye disease in these regions because of dust blowing about.

We developed our plan: three Wastewater Gardens, one serving two of the houses and a workshop on the lower side of the community, another two for the other houses. We'd make sunken gardens so that sewage would flow by gravity whenever possible. This would limit reliance on pumps.

We decided the septic tanks would have to be replaced. New septic tanks would be placed as high as possible so as to have the maximum slope possible for the gravity feed to the gardens. To do that meant we had to separate kitchen graywater which came from the front of the houses, from the toilet and shower water that came from the back. It's the drop from toilets to septic tank, generally required to be at least 2% (e.g., a slope of 2 inches per every 100 inches) that ensures that solids will flow reliably. Other wastewater only requires a 1% slope since it is mostly liquid. One house was high enough that we could safely install a normal Wastewater Garden and gravity leach drain.

The situation on the low side of the community was trickier. There was standing water and a natural wetland a stone's throw behind the houses there. Our solution achieved gravity flow from the houses to all the Wastewater Gardens. But to discharge the treated water from the control box in two of the systems, there would be submersible pumps to help during the wet season by sending the treated wastewater to leach drains located on higher, drier soils in the Central Park. The total cost of the Wastewater Garden project construction was a fifth of a sewage lagoon.

Klaus Eiberle (left) and me. Klaus helped construct and plant the Wastewater Gardens at Emu Creek Community.

Flowers, Dogfights, and Stones

The systems went in with little mishap. The community was pleased with the variety of plants. We included coconuts, papaya, and banana trees as well as canna lilies and heliconia and other shrubs. We didn't use oleanders since their leaves are poisonous and that's a concern for the safety of children in aboriginal communities.

We worked to make the gardens as culturally acceptable as possible. Robyn was a great diplomat. She had grown up in the outback. In 1995 she was honored as the Kimberley, Western Australian and Australian "Rural Woman of the Year" in recognition of her ecological work at Birdwood Downs. She knew bush culture enough to not put up with any *humbuggery* (bullshit). When anyone began complaining about the houses at Emu Creek provided by the government, Robyn would stop the *whingeing* by relating how she'd

Dreamtime stories, painted by resident artists on the control box covers of the Wastewater Gardens at Emu Creek community, Kununurra, Western Australia. At bottom, Ned Johns shows his painting of the snake chasing the emu, the story which gives the area its name.

grown up in a house without any running water. She and I came up with the idea of turning the control boxes into art displays. Why not get resident artists to put Dreamtime paintings on the control box covers? These were their Wastewater Gardens, not ours. It was agreed. Soon, adorning our three control boxes was a painting with a snake dreaming, another with long-necked turtles, and a third telling the Dreamtime story of a snake chasing an emu. We trained a number of local health workers how to monitor and maintain the systems. Everyone in the community was happy. They loved seeing white government workers come out to Emu Creek to look at their new gardens.

Of course, there were some glitches and more learning ahead. First there were children playing in the gardens. So, we conspired with Clayton Bell and Greer Ashby, who were the local Health Department and Shire Environmental Health officers, to spread rumors that there were snakes in the system to scare the kids away. It didn't work. The kids weren't causing any damage, and there was no evidence they were digging in the gardens or coming into contact with any wastewater. But the Health Department was concerned. Next, kids were seen taking rocks and throwing them at each other. That really got the bureaucrats going. It seemed useless to point out that there were far more dangerous things lying around, like old steel or wooden fence pickets—Australian wood is really hard—and that kids are kids.

Then there were occasional crises to do with adjusting the float switches correctly. We finally got the plumbers to go and fiddle with them, since we were hundreds of miles away and only coming to Emu Creek every three months to do water quality tests and check on the systems. Despite the fact that we had an agreement with the local resource agency and had also trained a number of local people, maintenance wasn't being done reliably.

The largest Wastewater Garden at Emu Creek after a year's growth.

I finally eased up on my resistance to fencing (seen as a solution to rock-throwing kids playing too close by) and agreed to let wire mesh fences, with a locked garden gate, go up around the Wastewater Gardens. That calmed everyone back down in Perth, at the Health Department.

Gulgagulganeng (Emu Creek) was a challenge, both technically and because of the social problems there. It's sad but the community was incapable of doing even simple maintenance tasks. The original plan (which called for having people from town who were supposed to help the communities) fell apart because of job turnover. In retrospect, the problems during the first decade of operation have not been overly serious and we have certainly learned since how to make our systems even less demanding. Importantly, we learned a lot about how to deal with issues like high water tables, which is a widespread concern in the Australian north during the wet season. News about the beautiful new Wastewater Gardens at Emu Creek inevitably spread far and wide, by means of the "bush telegraph."

The view after chain link fencing was put around the Wastewater Gardens, Emu Creek community.

Newly planted Wastewater Garden at Coco Eco, Coconut Wells, Western Australia. The family house is to the rear, the steps and deck lead to the tourist rooms.

Coconut Wells and the Phantom Vent

The Kimberley is Australia's last frontier. Our property, Birdwood Downs, is near one end of the 500-mile-long (800 kilometers) Gibb River Road, a famed dirt road playground for four-wheel-drive vehicles. The road winds past magnificent fossilized coral reefs, wild rivers, and gorges and waterfalls, amid glorious tropical savannah landscapes. One hundred fifty miles (220 km) to the south is the town of Broome, once the pearling capital of the world. It has a unique culture and world-class sandy beaches. When we first started our work at Birdwood Downs, tourists were as scarce as hen's teeth. But the beauty of the region is becoming known, and Broome is now being marketed as Western Australia's answer to Queensland's Gold Coast and Surfer's Paradise. We've seen what this impending development can do to a place from experience in the Yucatan and in Bali. Broome has now grown to 15,000 full-time residents, and 40,000 visitors at the peak of the tourist season. When we suggested a demonstration project in the Broome area in 2004, it seemed good timing on our part. The Coconut Wells Eco Bed and Breakfast is a twenty-five-minute drive from Broome, at a beach that's much less crowded but as stunning as fabled Cable Beach.

The Coco Eco has magnificent open-air architecture and a solar-powered electricity supply. Set back from the beach on sand dunes that give houses in the area excellent views, the family house and three guest rooms are connected by raised walkways. We were elated when we realized that aesthetic considerations and plumbing needs meant that the Wastewater Garden would best fit in the central courtyard.

The Wastewater Garden was planted during the wet season, which meant there was extra cloud cover and rain, which was easier on new plants. It has a few large foxtail palms, which supplemented the plants we'd grown in the nursery at Birdwood Downs. All seemed well until a couple of weeks later when we got a phone call from one of the owners at Coco Eco complaining about a faint but persistent smell. Brad Riley (who helped on this construction and also Joy Springs) and I went over the checklist. Was the septic tank well sealed? Yes. Any exposed wastewater in the garden? No. Was there an airtight cover on the control box? Yes.

The beautiful sandstone rocks to cover the geomembrane liner were collected on the Coco Eco property. Hauling and placing rock gave a chance for us to glory as Stone Age men.

It was true. There was a faint sewage smell. How could that be? The wastewater was well covered with gravel. We put more gravel on top of the inlet pipe from where the smell seemed to be emanating. No improvement. Finally, we called the plumber who was doing the installation, to ask him what he thought. He told us he hadn't yet installed a vent at the main house, since the plumbing to the guest rooms wasn't complete. We, on the other hand, had completed the Wastewater Garden early so that the owner's family could move into the main house. Aha! The system was surely venting through the inlet pipe. The plumber suggested we open up another pipe to serve as a vent. There was still some odor, but before we left, it occurred to me to also raise the water level in the Wastewater Garden so that there wouldn't be any airspace left in the inlet pipe. No air in the pipe meant there would be no fumes from the septic tank. It worked like a charm. The problem would have never been noticed if the septic tank had simply drained out to a leach drain as those pipes are well below the ground. I now understood that those air vents installed by plumbers around the world—take a look at any house roof and you'll see them—have a real function!

The Wastewater Garden at Coconut Wells Eco Bed and Breakfast is still flourishing. I laughed when I received a phone call from several colleagues who'd gone to visit and walked right by the courtyard, perplexed because they couldn't find the system. They mentioned the lovely landscaping they'd seen as they looked for the Wastewater Garden, not realizing that the courtyard garden itself *was* the system.

Joy Springs: Inverted Leach Drains and Butyl Tape

The community of Joy Springs (Eight Mile) is fifteen miles east of Fitzroy Crossing, in the heart of the West Kimberley. In contrast to Emu Creek, it is a well-managed community. Joy Springs's inhabitants are basically members of an extended family, the descendants of Jock Shandley, the longtime head stockman at the nearby Go Go Cattle "Station" (Aussie for ranch). Jock is said to have been the first aborigine in the Kimberley to own a car or get married "European-style." Until a national referendum in 1967, aborigines were not full citizens of Australia. They weren't allowed to vote, needed permits to buy alcohol, and were not paid wages. In the outback there were large aboriginal camps on all the cattle *stations,* where they were given shelter and supplied with food in exchange for work. The men worked as *ringers* (cowboys), and the women did mainly domestic work in the white-owned homesteads. After 1967, aborigines became Australian citizens, legally entitled to the same wages as white workers. But an unforeseen consequence of this was

One of the long and narrow Wastewater Gardens at Joy Springs community, Fitzroy Crossing, Western Australia. We had to reuse leach drains that were dynamited through the hard subsurface rock layer.

that stations could no longer afford to keep all the aborigines living there on the payroll. Most left to live in nearby towns, where they received welfare payments, and with little to do with the money, alcoholism became rife. In recent decades there has been an effort to resettle aborigines on their ancestral lands by buying pastoral leases and creating outstations for them. A pioneering family, the Emmanuels, who owned stations in the Kimberley, including the Go Go, gave Jock a square mile of land for him and his family. The Joy Springs community is dry—no alcohol allowed. The place is well-kept, and its members are quite industrious, improving it all the time, building new sheds and fences and doing landscaping.

Having heard about our work at Emu Creek, people from Joy Spring approached us. At Joy Springs, gravity leach drains at a number of houses on the low-lying end of the community failed when there was a big wet. Some of the leach drains were also severely damaged because cars drove over them. We took some measurements, looked at possible solutions, and explained the new systems to the com-

One of the WWG systems at Joy Springs before planting. We had to make sunken gardens to be able to use gravity flow without pumps, to get the septic tank discharge to the gardens.

We hired several young people from the local indigenous communities to help with installation. Here, at Emu Creek, we relax leaning on our shovels.

munity. The community was eager for the greenery the Wastewater Gardens would bring. Funding was supplied by the West Australia Department of Housing. Our major headache, it seemed, was going to be rock.

The ground under Joy Springs is a hard sheet of rock. The old leach drains had been installed fifteen years earlier, using dynamite. A backhoe wouldn't be able to dig through the rock. The problem was, nobody knew exactly where those leach drains were or how deep they were.

We developed our plans accordingly. For the problem houses we designed long, narrow Wastewater Gardens, narrow enough to fit in the old leach drains. We'd have to make "inverted leach drains" for wet season disposal of treated wastewater. They're called that because instead of being low in the ground, they are raised up and have perforated pipe laid under the soil so that sewage water can be dispersed into soil well above the groundwater level. We planted the inverted leach drains with a mix of native plants, fruit trees, and flowering shrubs. We installed drip irrigation lines and automatic timers so that the gardens would receive water year round. When the wet came, the inverted leach drain plants would benefit from the extra water and nutrients in the treated wastewater being pumped up to them.

Doing a project near Fitzroy Crossing taught us about the difficulties of obtaining materials and equipment in the "back of the back of the beyond," the "never-never," "beyond the black stump"— colorful expressions Australians use to describe life in the outback. We ran into Murphy's Law on this one: Everything that could go wrong did go wrong. Equipment we were counting on renting was either vandalized, bogged down because of the wet, or was damaged by incompetent operators. A nearby mine had promised us materials for the project as a goodwill gesture, but it was now in receivership and couldn't deliver anything. The gravel the mine promised us was ideal for building inverted leach drains but was too dirty for the Wastewater Gardens. Clean gravel was impossible to get. To bring a large truck from Derby would cost thousands of dollars. So, we arranged to get local river rock, some of which was unfortunately delivered mixed with sand and had to be returned. Nevertheless, with patience and persistence, the systems finally went in.

But we had a slight problem ... that is, a problem I desperately hoped would be slight. One of the Wastewater Gardens had a leak in its liner. The system had passed its "leak test" and it was holding water. Only later did we notice a drop in the water level. I suspected the leak was caused by a sharp rock that had somehow punctured the liner while filling it with gravel.

Although backhoes and front-end loaders were used, old-fashioned shovel work was also necessary. Brad Riley (in rear) and I are excavating leach drains at Emu Creek prior to installation of perforated piping. The white box, top center, is the control box of the Wastewater Garden which will feed treated wastewater into the final subsurface irrigation leach drains.

The Perth geomembrane liner company had repeatedly advised us that the only way to seal our liner of low-density polyethylene (LDPE) is by heat-welding it, to fuse together two sheets of the material. This is one of the reasons we've now switched to using EPDM, a sturdier geomembrane liner that's easily glued and repaired. Out in the bush it was impossible to do that kind of precise heat welding. Imagine then my delight when I spoke to one of the company's technicians about our "slight problem" and was finally told that double-sided butyl tape can patch polyethylene.

I was thrilled but I was also livid. Why hadn't we been told about this repair method before? Had we known, we could have saved time and money. Was this butyl tape a new product? I called my contact again. How long would a butyl tape patch last? The answer was swift: "Butyl tape patches have been done on water dams in Perth and have held for years and years." What does one conclude from this? Do technical people only tell you what they feel like telling you? Or, if you don't ask, they won't tell?

It was time to play Sherlock Holmes and find the leak, now that we had a way to fix it. Of course that meant shoveling a few hundred pounds of gravel, ever so carefully, away from the suspect side so as to expose 5 to 6 feet (2 meters) of the liner. We're normally very protective of the liner, so we took our shoes off when standing on the liner as the gravel was added. We also checked our pockets for knives or screwdrivers and never used metal shovels anywhere near the liner. We found the hole. A sharp piece of rock was jutting out a quarter of an inch through the liner at a point almost exactly 6 inches (15 centimeters) below the waterline. Within five minutes the protruding rock was removed, and the liner was sealed with the magic butyl rubber tape. Then we filled the system with water. The liner passed the leak test. There was joy indeed at Joy Springs.

The Paradigms, They Are A-Changin':
A Cure for Diaper Phobia

"It's an ill wind that blows no good."
– William Shakespeare

In Perth, a mere 1,500 miles (2,400 kilometers) south of the Kimberley, there has been a serious drought for years, some say for more than a decade. Water use restrictions are in place in a city that has three-quarters of the population of Western Australia. Restrictions on graywater recycling are really unpopular, especially when people see their trees and gardens withering while their water bills go up. Such unrest among its citizens has now helped the State Government to reverse its longstanding policies and not only to permit but even offer subsidies and rebates to encourage homeowners to install graywater recycling systems.

What a revolution in thinking! A decade earlier, the official position was that graywater recycling was dangerous and illegal—another example of why we should never give up hope that old ways will change, that reason and common sense will prevail. The change that happened in Western Australia is occurring around the world. And this is very important, as graywater recycling is another of the emerging technologies helping to transform sewage from the pariah toxic waste into its rightful status as a renewable resource. One of the chief obstacles to graywater recycling has been "diaper phobia," or as it should be clinically known (tongue in cheek), "nappy neurosis." Why this is relevant to our investigation of the world of shit is that this condition has historically afflicted a large number of the bureaucrats who issue permits for the treatment, disposal, and reuse of wastewater.

I had first glimpsed this phobia when I was toured through the "Casa del Agua" (the Water House), a demonstration project by the University of Arizona's Office of Arid Lands Studies and the City of Tucson. This demonstration house showcased rainwater collection from the roofs, water-efficient fixtures, the use of desert plants for landscaping (to minimize water needs), graywater recycling into irrigation, and a small constructed wetland for graywater treatment. But when we quizzed the city official making the presentation, he admitted that graywater recycling was forbidden under existing Tucson water use regulations. Under questioning from the evermore skeptical graduate students, he further admitted that he personally ran his laundry water out to his backyard for irrigation, and ventured the opinion that maybe a majority of the people in Tucson did likewise. But it was against the law because of the lurking menace of DIA-PERS! In the years since, graywater recycling has also been permitted in Tucson, but here is my perhaps overly vivid impression of what I have gleaned (but what was not overtly said) over the years from assorted health officials:

"Well, you see, babies' nappies are full of **doo-doo,** which contains microbes, some of them potentially carrying deadly and infectious **diseases**, and these are not all killed in the hot water of a laundry machine, and if you allow innocent, incompetent **untrained people** to take this graywater when it leaves the laundry machine and use it to water plants, and if those plants happen to include leafy vegetables which make their way unwashed to your **dinner plate**, and they get eaten… or if someone touches the dripper or hose and gets some of those deadly **germs** on their hands and wipes their **mouth** before washing their hands… and if those **dangerous** germs at large get past those people's immune systems in sufficient numbers… and it's seen that I, in my official capacity, signed off on anything which abetted and assisted making those germs' presence in our midst more likely and legal, and people get **sick and die**, there will be **nasty** newspaper and TV stories, **and I**, a good, careful civil servant, will be **blamed** for my foolishness in changing any regulation, no matter how outdated, and doing anything that increases the hazard of anything bad happening. So, would you **stop bugging me** about reusing that deadly nappy-infested laundry graywater and just **shut up** and continue to send it down to your septic tank or the city sewer, where real professional, skilled **technicians** can take care of it, and don't forget to **pay** your sewerage fees and your ever-escalating-in-cost water bills on time, or we'll turn off the tap and you and your damn garden that you care too much about anyway can **dry up and die** … but it will be **your fault**, not mine, got it?"

A textbook case of diaper paranoia, though outbreaks seem to be diminishing in health officials. For those interested in some reality thinking on this explosive topic, a few facts: Your baby's germs are not likely to be the Ebola virus or typhoid. Loving parents are getting those cute baby germs anyway from all the hugging and kissing that accompanies the happy days of parenting a small child (and changing countless diapers). Yes, accidents occur, so other clothes in that washing machine may have a touch of doo-doo! As of now, there has not been a single recorded instance of a serious disease being transmitted through the limited legal or more widespread people-instigated graywater recycling/reuse for irrigation that's happening all over the US and elsewhere in the world. Most graywater is simply soapy water, and the soaps and detergents add nutrients that make the water better for irrigating your plants. This is spreading so fast that there's now a word—fertigation—for irrigation enhanced with the nutrients (organic fertilizers) of graywater or treated wastewater. When you do put in a graywater recycle system, there are soaps and detergent types that are better for garden soils and plants—those lower in salts for example.

Clean, fresh water is great for drinking but is pretty much a nutritional zero for plants, apart from the moisture. One might also add that making graywater reuse legal and setting up useful and practical ways for the homeowner to accomplish using it might be more in the public's health interests than protecting us from diaper germs by keeping graywater recycling illegal. Graywater recycling would also reduce the load on septic tanks and sewage treatment plants, which should just deal with toilet and more concentrated wastewater. Graywater recycling also conserves enormous amounts of freshwater. In the United States, for example, it is estimated that in dry regions, from 50-70% of the clean, potable water supplied to houses is used for irrigation, and lower percentages are used for this purpose elsewhere. Keeping graywater reuse illegal means that all that landscape irrigation is using our planet's precious, finite supply of *potable* freshwater—our drinking water!

While We are Changing, Let There Also be Green in the Leach Drains!

The paradigms are also happily changing with regard to plants in leach drains. The old attitude was "keep those plants out, their roots might clog the pipes!" Now, as with graywater recycling, there is appreciation that having shrubs and trees in the leach drains will result in greater utilization of the wastewater and nutrients, preventing them from polluting water sources or going to waste. The reasonable precaution is that no plants having invasive root systems be used.

With each new Wastewater Garden that goes in, regulators are also getting a bit more relaxed, realizing that constructed wetlands technology is a public health benefit rather than a health hazard. We received a series of Australian Government Community Water Grants, as well as funding from the W.A. Department of Housing, to put in five gravity-flow Wastewater Garden systems at Pandanas Park, another aboriginal community, along the Fitzroy River outside Derby. Australia takes its water shortages very seriously and is preparing should global climate change increase droughts. Our funding came about in part because of all the water-saving we achieved by creating green areas, using wastewater.

It seems light-years ago since I was contemplating starting a prison journal because of an unapproved flower garden. Many challenges remain, such as installing Wastewater Gardens in really remote areas like the aboriginal outstations on the Dampier Peninsula, where word has spread about our work. We continue collecting native plants from the Kimberley for Wastewater Gardens eventually being installed at Conservation and Land Management (CALM) areas and in national parks, and we are promoting the technology across Australia.

As we had done in Mexico and elsewhere, we began our Australian work under the auspices of PCRF, for which I served as vice president for several years. In 2004, PCRF ceased operation of the Wastewater Gardens division to concentrate their efforts on other projects. That's when I founded Wastewater Gardens International, continuing to use the Wastewater Gardens® trademark with permission. PCRF has since resumed implementing Wastewater Garden systems through the Biosphere Foundation.

Wastewater Gardens International has a regional representative, Andrew Hemsley, who heads up Integrated Natural Systems, looking at applications all over Western Australia, including in the Perth area and the southwest of the state. We also continue working with Birdwood Downs as the local contractor for Kimberley projects.

Wastewater Garden at the Tirtagangga Royal Water Gardens, Bali, Indonesia

10
Travelin' Man: On the Scent of Solutions around the World

"In the land where no one knows you, there do what likes you."
—Marouf the Cobbler, *One Thousand and One Nights.*

Bali: Island of Gods and Goddesses

A green jewel of an island, Bali has an ancient yet vibrant Hindu culture, in a mountainous land sculpted by rice terraces. The grace and beauty of its people is startling. They are always smiling. I learned that since they believe in reincarnation, their deepest wish, assuming their karma is good, is to be reborn Balinese.

I had been enchanted by Bali when I passed through too briefly in 1982. In 1994, Bali hosted an international conference on bamboo, and I decided to present a paper at this scientific meeting so I could find out more about Balinese culture. I was also very interested in bamboo because I thought we could use it in our constructed wetlands.

At the conference I met Emerald Starr, an American who had moved to Bali from India. He is a gifted designer who had built several beautiful houses in Bali. He and his partners were planning to build an upscale hotel and retreat center in northeast Bali, just below the island's most sacred mountain, Mount Agung. He wanted to build the project in an ecologically sensitive way, entirely out of bamboo, to show that precious tropical hardwood trees need not be used. Bamboo is an astonishingly versatile material, and harvesting bamboo actually improves the health of the remaining stand. It has been referred to as "vegetable steel" because of its strength and flexibility. There are more than 1,500 species worldwide, and more than a billion people live in bamboo houses. One of the fastest-growing plants in the world, some types of bamboo grow to more than a hundred feet tall, eight inches in diameter. A young sprout can reach full height in a few weeks. It then slowly hardens and becomes incredibly durable. One species has been recorded growing over 3.5 feet (1.1 meter) in a single day. Almost 2 inches an hour!

Emerald Starr wanted to make Sacred Mountain Sanctuary, in Sidemen, Bali a shining example of ecologically sensitive construction. He was keen to hear about my work on constructed wetlands and wanted to incorporate our technologies into his retreat center design.

We decided to have graywater recycling for the bamboo villas' shower and sink water. Graywater would be used to irrigate areas of tropical flowers and trees in each villa's garden. Blackwater (from toilets) would be treated in Wastewater Gardens. Some Wastewater Gardens would treat two villas. Where there was enough slope, we connected a single Wastewater Garden to a cluster of villas. The largest Wastewater Garden was connected to the kitchen and restaurant.

We designed Sacred Mountain Sanctuary using graywater to irrigate the villas' gardens. Coming through drip-irrigation outlets, the graywater is more evenly distributed. Slow release of the water enables the plants to absorb it quickly, and so there are no pools of standing water around for anyone to touch.

The Wastewater Gardens were made with a concrete liner, as in Mexico. In Bali, labor is even cheaper. Skilled artisans make two to three dollars a day for construction work. These were the early days. We had just started our demonstration systems in Akumal, so I sent some drawings to Emerald along with lengthy e-mails explaining important issues concerning planning and construction.

Emerald Starr, environmentalist for over 30 years, received the Tri Hita Karana Environmental Award in 2000 for his work on Sacred Mountain Sanctuary, a retreat center made with bamboo buildings and Wastewater Gardens in central Bali. He also joined the Balinese Royal Family of Karangasem to restore the Tirtagangga Water Palace.

When I checked the first few systems I found that the gravel used was almost completely packed with sand and soil. This need for empty spaces (pores) between the gravel stones was lost on the workers (who didn't really know what they were building). Unless there are spaces, the system can't hold much water; the water fills them. The technical term for this is "porosity." For finely sieved small gravel, that's around 40% of the volume. Calculating how much water can be held determines how long the wastewater stays in a Wastewater Garden before being clean enough to be released. With gravel that's dirty, you may get a tropical garden, but you'll get very little treatment because the wastewater won't stay in the wetland long enough before it's displaced by new wastewater flowing in.

What could we do to fix the situation? I groaned when I heard how far from the site the gravel had been trucked. But then, the realities of Indonesian economics kicked in. The gravel was taken out, shovel by shovel, and screened right next to the system by two workers holding a wire mesh frame. Even if it took a couple of days per system, that correction added only a few dollars to the cost of construction!

Emerald has great talent as a landscaper and designer of living spaces. There is a tradition in Bali and elsewhere in Asia of locating bathrooms and showers outside, set in a lovely mini-garden. The term for bathroom in Bahasa (the dominant Indonesian language) is "kamar kecil"—small room. What a refreshing change it was to have bathrooms in a tropical garden instead! Clever use of half-height walls hid the toilet corner, and we took showers surrounded by lovely plants. The walls blocked any view from outside a villa. One touch I loved was that in some of the villas one could actually see Wastewater Gardens from the toilet. One could watch flowers while sending nutrients their way. This is seeing the "end in the beginning, and the beginning in the end" as a Middle Eastern proverb tells us.

The Wastewater Garden systems
for the kitchen/restaurant at
Sacred Mountain Sanctuary, Bali.

Yes, We Have No … Data

Sacred Mountain Sanctuary was completed in 1997. We now
had a demonstration project in Bali, with seven Wastewater Gardens
but I was looking for a reason to return to the island I loved, so
Emerald and I decided to start a business.

Easier said than done! We visited many potential customers.
In the process, I found myself doing a postgraduate course in the
treatment and mistreatment of shit. One memorable lesson was at
one of the most expensive and exclusive resorts in Bali where a mere
$1,200 bought you a private villa for a night. Not bad value in a
country where half the people earn three dollars a day!

Emerald was a friend of the hotel manager and had heard that
the resort was having problems with its sewage treatment plant. As
we made our way down steep slopes toward the river gorge that the
resort overlooked, the smell grew more and more pungent. Finally
we arrived at the treatment plant, which had been in operation for
eight years but was now starting to leak. Its steel tanks were corroded.

We went back up the hill to see how the final disposal of the sewage was done, and I could scarcely believe my eyes. At the top of the hill there were some rice paddies that had been converted to vegetable growing. Sewage was flowing along the channels meant to irrigate the restaurant's kitchen gardens, including its leafy greens. We had just been told that the water used for irrigation was *not* being disinfected beforehand. We warned the hotel manager of the potential disaster he was facing. Although they did not upgrade to a constructed wetland treatment system, at least the resort immediately stopped irrigating vegetable crops within a few weeks of our visit. And this particular hotel and resort had won an award a few years back as the best hotel on the planet! Most of the visitors do not take the walks we did—out to the edge, to the sewage plant, or to the vegetable garden.

This was the start of a learning curve. Scratch the surface of this island paradise and you find some serious problems developing in Bali, much like what I'd had seen in the Yucatan peninsula. The island of Bali was also now a major tourist destination with as many visitors per annum as the resident population. But the visitors use five to ten times more water per person per day than the Balinese to flush toilets, take hot or cold showers, use swimming pools, Jacuzzis, and so on. In the rush for the tourist dollar, sewage treatment was not a major concern. We visited Kuta Beach, another tourist area, and learned that many hotels ran their waste pipes under the sand, straight out to the ocean! Those who had their septic tanks pumped out sent the waste out on trucks, which simply dumped it out of sight, in the mangroves or wherever, just like in the Yucatan. A new sludge-drying treatment plant was so badly designed, and the smell emanating from it so horrific that it was closed down due to complaints by irate neighbors, after operating for just a few weeks. Who knew how much of the original budget was actually used to build the facility? I had heard that it was routine for 50-80% of the budget to be skimmed off by corrupt government or military officials.

The problems are not limited to those created by visitors. Balinese use their extensive water networks for rice paddies, personal hygiene, swimming, and toilet facilities. But this water gets more and more polluted as it flows downward to the coast. Balinese house compounds, so beautifully architected, each with a private temple, have septic tanks but they are unlined, just as in Mexico. So the sewage pollutes the groundwater and rivers, as it makes its way to the coast.

I was back in Bali for the year 2000 millennium celebrations. Emerald and I gathered together friends and other people we had met, to start a non-profit foundation (Planetary Coral Reef Foundation—Indonesia) to promote Wastewater Gardens in Bali and elsewhere in Indonesia.

At seminars at Udayana University and at meetings with the government in Denpasar, Bali's capital city, I would get this strange *Alice in Wonderland* feeling after listening to presentations. Something wasn't quite right: I began to realize that I was hearing eloquent speeches but no one was actually presenting any data on water quality, sewage treatment, or river or shoreline pollution, or any statistics on the current health of children or adults. It was a chimera of arm-waving and fine-sounding phrases. I asked people why this was the case and was told the following cautionary tale. A few years before, someone had leaked to an Australian magazine data on ocean water quality that been collected by the World Health Organization. A story was published which stated that swimming at popular beaches in Bali could be dangerous because of truly alarming levels of bacteria and pathogens in the water. Tourism suffered as a result, but eventually people forgot the story. Thus, there was a natural reluctance to present truthful data.

An Australian scientist leading a World Bank environmental assessment program in Bali told me that it was quite common for data on rivers to be the same, year after year. He suggested two possible causes: Money for any research had been stolen, in the local "corrupsi" way, or laboratories that were supposed to do the analyses couldn't get their equipment to work, so they simply presented the same data as the year before.

Never mind real-time data. Often, direct observation is ignored. We had been told that there were problems with the sewage ponds built for the tourist "ghetto" (a ghetto of expensive five-star hotels) of Nusa Dua. A foreign engineering company had built a type of constructed wetland there that is no longer in favor since it has not performed as well as expected, and fared worse than other approaches. This type of system uses water hyacinth, a very fast-growing floating water plant. In Florida and elsewhere in the tropics, water hyacinth is invasive and clogs up the waterways, but that is another story.

The sewage treatment lagoons were built some time ago, and I was surprised to see them still in use. We'd heard that the water quality was very poor and that the lagoons overflowed into the adjoining mangroves by poor design, not on purpose, in the wet season. We went to have a look. On the day of our visit we saw a truck backed up against the side of one of the lagoons. Men had waded in and were catching fish, putting them into the back of the truck, thence to market.

The wastewater in the lagoon is not disinfected, so what the fish contain is anyone's guess. I was curious to follow the truck and find out where the fish were being sold, but we had an appointment with the director of the company that was managing the sewage treatment plant, so I couldn't. When we met the director and told him what we had seen, his response was very clear. Fishing in the lagoons is not permitted. There are security guards on duty to prevent just that.

Therefore, you are mistaken. Total denial! By the way, did anyone see where the White Rabbit went, the one late for an appointment?

Further travels around Bali and elsewhere in Indonesia deepened my sense of foreboding that our planet is in trouble. Anyone who has traveled in developing countries can see this, especially if you get away from tourist resorts catering to people from affluent nations. Bali is one of the more prosperous islands in the Indonesian archipelago, yet there is a sense of desperation among the populace. Some benefit from the influx of tourist dollars, but others are left behind.

We went to National Parks in northwest Bali, where almost all valuable timber in the forest had been cut down, in spite of the parks' protected status. Nyoman Sumartha, a local leading our expedition, was from a nearby village called Ekasari, and told us the issue is dividing his community. Some oppose the loss of the forest, out of love for Nature and the trees. They also fear environmental consequences affecting their water and causing soil erosion. Others argue, "Why let the big corporations or the corrupt military be the only ones to profit from illegal logging? We should do it too." To our friend's dismay, we came across a group of men cutting down one of the last teak trees. Nyoman recognized their leader as a schoolteacher who had once taught him at the local school.

In the capital, Denpasar, I saw young men as well as whole families living under tables at the batik factories where they worked. We went on an inspection tour, led by Ibu Kartini, a dynamic scientist who had pioneered a project composting animal manure from Denpasar stockyards. She was also raising earthworms. She was affectionately known as the "Earthworm Queen." She advocates the medicinal use of earthworms as well as their use as a nutritional supplement. I had met her at Udyana University in Denpasar. We were kindred spirits—immediately recognizing each other as members of the worldwide earthworm fan club!

Water used by the batik dye factories flows into to the irrigation channels of Denpasar. Animals drink it and people swim in it and collect water for home consumption. The polluted water then flows into rivers and finally to the ocean, the ultimate waste dump. Some cows grazing near the factories had recently died, and the government wanted to show it was doing something to prevent violence breaking out between batik factory workers and their neighbors. I was shocked by how poor Balinese and Javanese batik workers were. The arms of the young men who work in the factories were immersed in batik dyes all day long and were thus completely stained by whatever color was being used.

Trust in God, But Tie Your Camel First

The new millennium saw a spurt in business for Wastewater Gardens in Bali. Our enterprise looked as if it might actually take off. For example, we installed a beautiful system at the Sunrise School, an eco-friendly establishment. We also installed a system at a number of nearby villas. Jil Posner, a longtime friend who has lived in Bali for many years managing various recycling enterprises so fell in love with the idea of a Wastewater Garden that she abandoned plans for a goldfish pond in the courtyard of her new home, north of Ubud. An Indonesian environmental agency recruited us to do tests to see if we could actually improve the situation at the batik factories. Constructed wetlands could be an effective way of absorbing the heavy metals and other pollutants from the batik factories' wastewater.

But just when I thought we had everything organized, people whom we had trusted with construction and our sales and marketing completely betrayed us by means of a corrupt scheme designed to steal our intellectual capital and good name. I hired a well-known Javanese lawyer, Agus Samijaya, known for his anti-corruption

Wastewater Garden at Sunrise School, Kuta/Legian, Bali. This 220 ft² (20 m²) system treats blackwater from 60 students and 15 staff and is used as an outdoor classroom. This was the first system in which we used water lotus plants in addition to tropical plants that thrive in Indonesia.

Wastewater Garden built by Jil Posner for a family compound near Ubud, Bali.

efforts. Pak (Bahasa for "father" or "sir") Agus drew up an impressive legal document detailing the many incidents of fraud, intellectual misrepresentation committed by our ex-partners. He scared the living daylights out of anyone, including government officials, who had agreements with our ex-partners. He made them understand too that they might be liable. Pak Agus sent documents to the media and government officials in Denpasar and Jakarta, Indonesia's capital. Officials with whom we had already been dealing now asked to meet with me personally, to draw up "new" agreements. Interestingly, my ex-partners vanished from the scene. I had learned a painful lesson: Betrayal is a far more unpleasant kind of shit than the good stuff we were used to.

Business as Usual

We started again. To promote Wastewater Gardens we linked up with Yayasan IDEP (Indonesian Foundation for Development and Education in Permaculture), founded by Petra Schneider, a Canadian/ Indonesian who had been living in Bali for more than twenty years.

Wastewater Gardens can be creatively shaped to fit available space and add landscape beauty. Here a lovely curving system. built for a complex of villas in Bali.

She had built an organization involved in key issues, from women's health to sustainable farming to waste treatment. At the time, she was promoting healthy kitchen gardens while also trying to help Bali recover from the trauma of the terrorist bomb attack that killed 202 people and badly injured 240 more.

Meanwhile, Emerald helped train some locals to take over the production of system blueprints, construction, and planting. Frank Wilson, a level-headed plumber and sewage engineer from Melbourne, Australia, now meditation teacher and management consultant, was another gifted individual who brought his expertise to our fledgling enterprise.

An invitation came from the Seacology Foundation to Planetary Coral Reef Foundation, which then was operating my Wastewater Garden division, to submit a grant proposal for a village-based initiative. We chose the Tirtagangga Royal Water Palace, in eastern Bali. The eruption of Mount Agung back in 1964 had damaged many statues in the gardens, and a restoration effort was finally underway. We received the Seacology grant and also key supplemental funding from the Livingry Foundation, and put Wastewater Gardens at Tirtagangga to protect the bathers and the nearby coast. Tirtagangga means "Holy Waters of the Ganges," and the gardens are fed from springs up in the hills. Two beautiful Wastewater Gardens have now added to the beauty of the Water Palace, a favorite spot for both locals and tourists. One Wastewater Garden treats the flush-toilet bathrooms and the other the Asian-style toilets.

Manado, Mangroves, and Bamboo

Through IDEP I met Ben Brown, the Southeast Asian representative of the Mangrove Action Project (MAP), an American foundation dedicated to mangrove conservation. Ben had worked as a teacher in Indonesian villages, especially in the Manado region of North Sulawesi (a large island north of Bali). Fluent in Bahasa, he married an Indonesian woman, yet he was still a quintessential down-home Canadian.

IDEP and MAP would collaborate once again with Seacology Foundation support, on putting in a Wastewater Garden to treat the bathroom wastewater at a bamboo processing facility in Tiwoho. When Ben and I first met, we came up with the idea of using mangroves for this Wastewater Garden. Since mangroves actually grow better when there is fresh water available, we could use the Wastewater Gardens at Tiwoho as an educational prototype. The layout of the gardens would emulate the way Indonesian mangroves spread out from the seashore, back inland. Indonesia's mangrove ecosystems are some of the world's most diverse, with more than forty-five species, including many that have nutritional or medicinal value, such as the nypa palm. Our work, in January 2003, also led to Wastewater Gardens being installed at a number of Bunaken National Marine Park resorts. The original Tiwoho center is now a facility for the restoration of mangroves in north Sulawesi.

The Lotus in the Shit

At one point I was staying in a *lembung*, a traditional rice storage building, which also served as a guesthouse at the old IDEP headquarters. It was a rickety structure, and one could see the tracks of termites, intent on furthering the dilapidation. I used a ladder to reach the loft. Downstairs, a raised platform looked out over a lotus pond. Sitting there, sipping from a bottle of arak (palm liquor), I realized something about water lotuses. I know they are a major symbol in Buddhist mythology. Padmasambhava, a manifestation of the Buddha, is said to have been born on a water lotus. I had always lumped lotus together with water lilies, both lovely plants that float on water. Now I saw that the water lotus was different. There it was, rooted in the mud. If it could rise up through several feet of water, why then couldn't it protrude through a few inches of gravel in our wetland systems? And so, water lotuses joined the ever-growing list of plants.

I had learned from a visit to the museum in Cairo that the lotus and papyrus were symbols of the Upper and Lower Kingdoms of ancient Egypt. Papyrus is the plant from which the first paper was made and catalyzed humanity's cultural transmission capabilities. Water lotus is a symbol of personal self-realization and transformation. Both are Wastewater Garden plants. We had used papyri in the first Bali-constructed wetland systems at Sacred Mountain Sanctuary. "The jewel in the lotus" is a recurring Buddhist image. Now we were demonstrating "the lotus in (and from) the shit."

The Bahamas: A Reality Check

After completing a number of Wastewater Garden projects in Bali, I was traveling a great deal—not that I hadn't traveled a fair bit before, but I had now discovered round-the-world (RTW) airline

Wastewater Gardens for Cape Eleuthera Island School in the Bahamas; this photograph was taken only a few months after planting.

tickets. The call for Wastewater Gardens was growing. It was especially interesting to go to places where no constructed wetlands had yet been built. We could be pioneers and, in the process, learn how different cultures dealt with their shit.

I offer a few of the scenes and stories that particularly struck me during these travels, not as an average tourist but rather a journeyman in the strange orbit of the world's wastes.

It came to pass that, after much back and forth communication, I found myself sitting at a table at the Cape Eleuthera Island School, in the Bahamas. The school serves gifted high school and prep school students, and it has a superb environmental curriculum. Jack Kenworthy, who was typical of the school's bright and creative faculty members, had invited me. Jack had been a student at Biosphere 2 when Columbia University was managing the facility in the late 1990s, and he had heard about my work.

Jack Kenworthy was convinced that installing Wastewater Gardens at the school would be a great way to show its commitment to sustainable technologies. Eleuthera is a relatively quiet and undeveloped Bahamian island and the Island School is situated on a pristine piece of land that juts out into the beautiful blue-green waters of the shallow Great Bahamas Bank. It is surrounded by mangroves, and there is a coral reef a short distance from the beach. The soil is predominantly limestone as in the Yucatan but is much sandier.

Also at the table was "Big John," a local who had done most of the building work at the Island School. I had been describing the finer points of a constructed wetland. Just as I was explaining that the purpose of a septic tank filter is to prevent solids from passing into the Wastewater Gardens, Big John exclaimed, "But *mon*, that's not gonna happen. Those tanks not got a bottom!" Jack looked stunned. He had assured me repeatedly that the septic tanks at the school were watertight, especially because I had told him of my experience in other parts of the world. Ordinarily, I'm not much into New Agey stuff like telepathy. But *this* was a moment of pure telepathy. We didn't need to speak. Both Jack and I were thinking about where those septic tanks were leaching out— through the sandy limestone back of the school buildings to the mangrove areas, where students routinely conducted surveys of mangrove and other coastal vegetation.

The septic tanks were sealed as quickly as Big John could organize things. Wastewater Gardens were installed in the center of the Island School campus. They had a beautiful layout of pathways around them. The gardens bloomed. Jack took a chance and planted coconut palms in the garden. An "expert" coconut grower in Mexico had assured me coconuts would not grow unless they had 12 inches (30 centimeters) of dry soil above the standing water. I had taken his word until I discovered just how wrong very impressive-looking books were about wetlands. I decided this expert might be full of it as well. The coconut palms thrived, growing even faster than usual. They weren't intimidated by an expert naysayer.

The Philippines: The Dead River and a Haven for Kids

Flying into Manila, capital of The Philippines, for the first time, as my plane turned toward the airport, I got a good view of the Pasig River, snaking its way through the city. It was a brown color, like slightly milky coffee. I had never before seen a river that color. I braced myself, not for the airplane landing but for the close encounter with reality that I knew was imminent.

Some friendly people who were sponsoring my visit met me at the airport. Manila is a pulsing megalopolis. The traffic was completely congested, and the streets were thick with crowds. The noise was impressive, even to an ex-New Yorker.

To save my client some money, I agreed to use a guestroom at a children's shelter rather than stay at a hotel. After recovering from jet lag, it was wonderful to wake up and walk through rooms filled with children and their caring teachers. I found out that the kids were there because they had been orphaned, abused, or abandoned. Some of the staff had seen our systems in Bali, and what brought me to the Philippines was to advise them on using Wastewater Gardens for a permanent home and school for the kids, to be called The Children's Village. I gave a talk about Biosphere 2 at the University of Asia and the Pacific, in Manila, focusing on ecological engineering and Wastewater Gardens in terms of living sustainably with the biosphere. At the lunch afterward, I was told the grim facts about the coffee-colored river. Manila, a city officially with eight million inhabitants (although fifteen million was probably closer to the truth if you included the never-ending urban sprawl), has *zero* sewage treatment. There was no sewage treatment in a city that ranked as one of the top five most populous on Earth! All of Manila's sewage goes into a river, which, as a result, is biologically dead. It is thus one the world's largest open sewers. It was no wonder there was interest in what we had been doing but, amazingly, no one could tell me of any constructed wetland projects in the country, although at the University of Asia and the Pacific, at least *some* people had heard of the technology.

I had a sudden flashback. Some years earlier, I had given a talk on more ecological approaches to wastewater treatment at the University of Tuxtla Gutiérrez, the capital of Chiapas in southern Mexico, at the invitation of a good friend and fellow graduate student, Hugo Guillén Trujillo. I hadn't traveled much in Mexico before giving that talk, so I wasn't that clued in to what was happening. After the talk, a few earnest young students came up to me. One asked if I knew what they did with their shit and wastewater in this city of several hundred thousand. I confessed I didn't and then listened in amazement as he described a collection system of pipes bringing the untreated sewage down to the river.

"Then what happens?" I asked.

"Nothing. It goes into the river."

At a loss, I could only blurt out, "But what do you do? Don't people use the river for swimming?"

"Of course we do," he replied, "but we only go into it upstream of our city." But of course, in this highly populated country, there are more villages and cities both upstream and downstream of Tuxtla, that add their own loads to what the river is carrying. The illogical, deadly conclusion of that way had now been shown to me in Manila—a brown, stinking, and very dead former river.

What kind of creatures are we who not only foul our own nests, but poison the water sources, rivers, lakes, and oceans of the biosphere that belong to all species?

On the outskirts of Manila, there is district after district where people live off and on top of man-made mountains of garbage. Here is a realm of purgatory far beyond anything Dante imagined in his *Inferno*. Garbage is sorted by type, and huge areas are covered by particular types of trash. The wretches eking out a living in this wasteland, retrieving anything of value, are similarly classified (plastic scavengers, glass scavengers, etc.). I had heard stories of disasters when the garbage sometimes spontaneously combusted or a makeshift village that was lost due to the sudden collapse of a veritable mountain range of detritus. The garbage ranges and the people living on them were almost beyond belief and brought me to tears.

The Carpathian Mountains:
Biosphere Reserves and Border Patrols

The East Carpathian Biosphere Reserve, in the Carpathian Mountains of Eastern Europe, straddles three countries: Poland, Slovakia, and Ukraine. Though these countries collaborated in its establishment, the area is not exactly user-friendly. In the middle of the Reserve, along the San River, which divides southeast Poland from Ukraine, there are border patrols. If you try to cross the river anywhere other than at a designated border post, you will be arrested as an illegal immigrant, regardless of whether you were simply hiking, riding, or bird watching.

I was hiking up a pristine mountain path with Andrzej Czech, a new friend who was working on restoring the European beaver to its native habitat. He was also interested in our constructed wetlands. Beavers are wetland builders, so they are sometimes blamed for floods because they build log dams across creeks and streams. The European beaver is almost extinct but is now being reintroduced as an important component of riverine ecosystems. Andrzej had just completed a PhD on beaver restoration in the Carpathian Mountains, at Jagiellonian University in Krakow, Poland.

The Carpathians Mountains of Poland hold bitter memories. This is an area where mutual hatred and ethnic conflict have simmered for centuries. Populations have been slaughtered by wars and genocide and moved around like so many chess pieces, as national boundaries were changed by politicians far away. After World War II, the Polish border with Germany was moved west by 100 miles, at the behest of the victorious Soviet Union, and a large part of eastern Poland became Soviet territory. Unwanted citizens were evicted from homes they had been in for generations. Andrzej showed me a place where a town of thousands had once been but now was marked only by a few headstones and some building foundations; the surrounding forest had reclaimed the area.

However, because of depopulation, the Carpathian region has escaped the ecological damage sustained in more developed areas. Andrzej and his colleagues at the Carpathian Heritage Society were now concerned that the recent opening up of the region to tourism be done responsibly, hence his interest in Wastewater Gardens. I was glad we had some experience installing constructed wetlands in cold climates because temperatures in the Carpathians frequently drop below -20 degrees centigrade (-4 degrees Fahrenheit) in the winter. Engineers design for the worst-case scenarios. Cold winters meant that we would need more area of wetland per resident than is required in warmer climates.

In theory, I knew constructed wetlands worked successfully in very cold climates. When I was working on my master's thesis I had come across a wealth of data, from the 1960s and 1970s, about growing forest trees using sewage. The research was stimulated by the search for alternative energy sources, like biomass, since this was the era of the first OPEC (Organization of Petroleum Exporting Countries) embargo of oil exports to the West. Some of the work focused on fast-growing trees as potential biomass producers. From that effort, scientists developed hybrid poplar trees that could grow 12 feet (4 meters) per year. Other research showed that sewage could be used in growing forest trees and that natural forests could safely receive wastewater.

There are natural wetlands in every corner of the world, in every biome, every type of soil, and every climatic zone. The Synergia Ranch Wastewater Garden covers around 200 square feet (20 square meters) and was planted with more than twenty-five varieties of cold-tolerant wetland grasses, shrubs, and trees, including a number of New Mexico native plants. At the Ranch, the first biomic demonstration project of the Institute of Ecotechnics, winter temperatures can drop as low as -10 or -20 degrees Fahrenheit (-25 to -30 degrees centigrade). So I knew we could design a cold-temperature-tolerant wetland.

We looked at a sewage treatment plant in Lutowiska, a small town of about 500 people, set in the middle of the Biosphere Reserve. In summertime, however, the town's population swells as tourists come to enjoy the spectacular surroundings and living traditions of the region. From outside, the treatment plant looked all right, but all too typical of high-tech sewage plants, it was hardly doing any actual treatment. Incoming wastewater and the water the facility released back out was virtually the same quality, and it was polluting a nearby stream. The government was threatening to fine the treatment plant. Because Biosphere Reserves are protected areas and "exotic" plants are not allowed to be introduced, we built Wastewater Gardens at Lutowiska and at a Jagiellonian University field station in the Magursky National Park, drawing from the wealth of native wetland plants.

Provence: Sculptures in the Garden

Aix-en-Provence, in the south of France, is home to the Mediterranean biome project affiliated with the Institute of Ecotechnics. Just north of the town is the *Domaine Les Marronniers*, which is used for workshops and conferences. Its seventeen acres (7 hectares) contain woodland, fields and orchards as well as artists' studios and a theater.

The *maison* dates from the seventeenth century and has more than two dozen rooms in its spacious confines. It was originally the house of a prosperous farmer.

People come to Provence to soak up its rich history, its stunning landscapes immortalized by artists like Cézanne and Van Gogh, and to enjoy its cuisine. What they don't come for is the plumbing. It is true that the nearby Roman-era aqueduct, the *Pont du Gard*, is an engineering and architectural marvel. Its original purpose was to carry water from a spring at Uzès to the Roman settlement at Nemausus (Nîmes). Les Marronniers' ancient Provençal plumbing using small diameter pipes, however, means that toilet paper is effectively a showstopper. Any material sent down the toilet that isn't readily soluble simply clogs the pipes.

The Wastewater Garden at Les Marronniers Conference Center, Aix-en-Provence, France, perfectly matches its Provençal setting. It is adjoined by a windbreak, typically used to protect farms and gardens from the *mistral*, the harsh wind which originates in the Alps.

Since the Institute of Ecotechnics acquired the property, its original beauty has been restored and enhanced. Provence is also home to the harsh wind known as the *mistral*, which roars down the Rhone Valley, from the north. It's known to drive people crazy, and it's hard on trees and gardens. That's why typical Provençal farms are shielded by robust lines of windbreak trees.

Cesco Rimondi, an Ethiopian/Italian and manager of the property with his wife, Molly ("Cyclone"), combines the pragmatism of a farmer with his visionary talent for fine art and sculpture. He built the Les Marronniers Wastewater Garden next to the vegetable garden and made it appear to be one of his art installations. He placed sculptures made out of wire, representing the human body, in the Wastewater Gardens. For me, the sculptures stand for the usual invisibility and taboo status our society puts on the subject of shit, that is, to ignore and look right through it. Here at Les Marronniers, the Wastewater Garden plants surround those bodies, growing on the "throughput" of materials that we eat to sustain ourselves.

Sally Silverstone picking flowers in the Wastewater Garden at Las Casas de la Selva, sustainable forestry project in the mountains of Patillas, Puerto Rico. This lush tropical system helps protect the nearby mountain stream from sewage pollution.

Puerto Rico Rainforest: Mud Eats the Machine

I was awakened by the unmistakable roar of the engine of a very large machine. Normally it's the *coquis* (frogs) that wake one up with the musical call that gives them their name, but on this occasion their cacophony, along with the morning chorus of birds, crickets, and cicadas, was no match for the sound of a backhoe. The engine noise increased, adding a note of urgency. I knew things were not going well. The backhoe had moved from a dirt track that led to the edge of a hillside next to the homestead where I was staying, almost to the bottom of the slope. It was now mired in four feet of mud. In vain, the man operating the machine tried to leverage it back up the hillside. It had rained heavily in the weeks before we came to Las Casas de la Selva to install the first Wastewater Garden in Puerto Rico. Las Casas de la Selva is an innovative rainforest enrichment and sustainable timber project. Annual rainfall in Puerto Rico can reach 120-150 inches (3,000-3,750 millimeters) in the mountains. The slope down to the site we'd chosen for the Wastewater Garden was steep. The system had to be built there because the septic tanks were only a little higher up the same slope. Below our selected site, the hillside dropped away sharply. A thick stand of mahoe, a Caribbean hardwood, had been planted to hold the soil. Trees grow quickly in the Puerto Rican rainforest, and the mahoes were already fifty feet tall, standing like sentinels, holding back potential mudslides that take whole hillsides away after torrential rains.

At the bottom of a slope, the Wastewater Garden at Las Casas de la Selva as it looked during planting in 2002. The system's final subsurface irrigation feeds the mahoe trees adjacent to the system.

After an hour of heroic but futile machinations with the backhoe, it was shut down, and the company that had dispatched was called out. It wasn't a local firm. I learned that local backhoe operators had come to the site but then politely withdrew, refusing to send their equipment anywhere near the rainforest mud. Ingenious solutions were now conceived. What about winching the backhoe up with a chain around the ceiba tree farther up the hill? The ceiba was about ten years old and perhaps twenty feet tall. It was still a child in terms of the giant tree it would eventually become. It was also, with its characteristic trunk and massive thorns, considered sacred. The Taino Indians of the Caribbean believed it was the connection between heaven and earth. The idea we might damage the ceiba or inadvertently pull it out of the ground, adding tree insult to machine-bogged injury, gave the grandstand generals pause. Cooler heads prevailed. So, we awaited the arrival of a bulldozer. It had a long tow chain and was able to pull the backhoe out of its mud bath.

William Dempster, a director of the Institute of Ecotechnics, stands next to the flourishing Wastewater Garden at Las Casas de la Selva. By 2009, a traveler's palm (a type of bird of paradise) has grown to over 25 feet tall in the wetland.

We were still left with two problems, though. The 500-square-foot (46 square meters) Wastewater Garden would now have to be dug out by hand, using picks and shovels. And forty cubic yards of gravel and rock would have to be hauled down that slope, by wheelbarrow. Fifty tons! Fortunately there was a large group of us that had come to volunteer time and labor. The Puerto Rico Department of Natural Resources was providing matching funds toward this installation of the technology. Mountain rainforest streams at Las Casas feed the reservoirs that supply water to the towns below. So, keeping them unpolluted while showing that beautiful, harvestable plants can be grown in the system would make their adoption by the *jíberos* (mountain people) more likely. Nine people, including three locals and me, took three days to do what a backhoe can do in four hours. Adding a couple more days of wheelbarrows working nonstop got us the gravel. It was time to open bottles of Puerto Rican rum and have a celebration. We had inserted a Wastewater Garden into the rainforest. Sky above, mud below, let the shit flow. Now we were ready.

Morocco: A Cascade of Wastewater Coming Down the Mountain

Beware of complacency. I thought I had seen everything, at least everything pertaining to the world of shit. I was wrong. I was up in the Atlas Mountains of Morocco, and what was coming down the mountain was not the kind of beautiful waterfall you see on the Discovery Channel.

I was traveling with other concerned parties: engineers from the town of Ifrane, Florence Cattin, our international projects director for Wastewater Gardens International, and a young researcher from Al Akhawayn University. The engineers from the city were professional-looking men and women armed with clipboards and printouts. They knew there was a problem. We had been tracking the course of the river that wound through the town. We were looking for the spot where the town's sewage joined it. The area around Ifrane is special since it contains the largest remaining cedar forest in North Africa. University and government officials were trying to get the area designated a World Heritage Site or Biosphere Reserve, to strengthen environmental protection in the region.

It was my first trip to Morocco. It was the first of May yet it had snowed heavily, followed by icy rain. The area around Ifrane had been described to me as the "Switzerland of Morocco," and it lived up to that. Indeed, Al Akhawayn University was designed by Swiss architects, and it sparkled like a Swiss mountain hamlet. It gets cold in the Atlas Mountains of Morocco. When told of the enormous daily water use by the students, many of them foreigners, I could not understand it. Finally, a sheepish explanation was offered: Dormitory rooms were not well heated, and students often ran hot water in the showers to heat the rooms.

An extensive network of pipes had been laid underground to carry away the university's waste. But the pipes didn't go to a treatment facility. Instead, the raw waste simply went down the mountain to join a maze of underground channels lower down.

We were now at the end of the line. Armed with their maps to the ancient buried city of shit, we were seeking not the source of the Nile but the outlet of all that crap. Everyone knew it joined the river somewhere, but where exactly? I had read research reports that downstream farmers were irrigating their fields with the untreated effluent, and the incidence of disease from contact with the sewage was a growing problem.

Eventually we found it, and the sight was enough to make us stop our vehicles. A cascade of filthy water was pouring down from the hill and then flowing along the road. From a distance it looked beautiful, but the engineers were alarmed: "There must be a blockage in the channel. This water is supposed to join the river farther down, carried by means of a large pipe. This does happen sometimes. The blockage has caused some of the wastewater to escape its designated track and it's now flowing freely overland." Indeed it was! The mixing of wastewater and river water was now complete. We looked in silence at the wastewater pouring out of

Aerial photo of Temacine, Algeria, showing the crescent moon-shaped Wastewater Garden flanked by date palm plantations. The old Ksar, caravanserai city, which is being restored, is on the right side of the photograph.

the pipe, into the river, and the engineers looked up from their site plans, now satisfied that they had a full grasp of the situation. A classic S-SNAFU. Shit Situation Normal: All Fucked Up.

Algeria and a Green Crescent Moon

In 2007 Florence Cattin and I designed and built the first constructed wetland for sewage treatment in Algeria. At first, what I was told about the area didn't seem to make sense. The area around Temacine in eastern Algeria is one of the breadbaskets of the country, with extensive date and fruit tree plantations. Yet it is in the Sahara desert, and one encounters extensive sand dunes before arriving at the green oases. How could too much water be a problem? Why had the French, who once ruled Algeria, built a canal that runs 100 miles (150 kilometers) to the Tunisian border, in order to get rid of unwanted water? When I first came to this region, in 2005, to attend a workshop, I began to learn about its strange ecological problems. Back in the day, date palms were watered with near-surface oasis water, but as more plantations were started, those water resources were soon insufficient. Date palms are one of the world's most water-demanding crops. Deeper and deeper wells had to be dug, and as the wells went deeper, thanks to modern technology they now go down some 900 meters (nearly 3,000 feet!). The water pumped to the surface is more and more saline and brackish. And as water evaporates in the desert heat, what remains becomes even saltier. Excessive use of water has also led to a rise in the water table, which is now in many places only three feet (1 meter) below the surface. So there's a Catch-22 situation: As more water is needed to flush salts from the date palms' roots, helping the dates survive, the water used is saltier and saltier, requiring yet *more* flushing water.

This situation is emblematic of the craziness of our modern world. Fossil fuels (used to power the pumps) are being used to

After the concrete liner has been completed, the system is filled up with water and left overnight to ensure it is sealed.

access fossil waters (deriving from ancient times when the Sahara was not a desert) deep underground, and both of them are non-renewable resources.

The drainage canal, which looks like a river, once hosted a rich variety of fish. It is now nearly lifeless because of ever-increasing salinity. Add untreated sewage from a growing population and you have a crisis.

Rachid Koraichi, a gifted Algerian sculptor and artist, invited us to come to Algeria to see if we could help. Rachid, founder of the Association Shams, was close friends with the Sheikh (Elder) of the Tedjani (Tijanyya) Sufi Order, a spiritual sect, one of whose main centers is in the town of Temacine. The Tedjani are very progressive, and their Sheikh wanted to bring innovations to the region, including Internet access, libraries, education for women, and appropriate modern technologies like Wastewater Gardens.

So, in 2007 we had the excitement of being wetland pioneers in Algeria. Our system, some 400 square meters (4,300 square feet),

Florence Cattin shows Sheikh Mohamed Laïd Tidjani, head of the Tedjani Sufi order in Temacine, the the progress of the crescent moon. In the background is the restored mosque of the Ksar caravanserai city.

Florence Cattin (center) with
Lamine Hafouda from INRA
Touggourt, Maurice Levy (to
my right), *Earth and Water*,
Portugal, with the owner and
helper during our visit to their
date palm plantation.

is designed to treat the effluent produced by up to 150 people, including the residents of an ancient Ksar caravanserai that is being restored. Our work was financed by the Algerian Ministry of Water Resources and the Department of Sanitation and Environmental Protection, along with the town of Temacine, with the support of the Belgian Technical Cooperation (BTC), a development agency that underwrote the cost of workshops we conducted for local engineers and government officials. With the help of enthusiastic scientists at L'Institut National de la Recherche Agronomique (INRA), we chose two dozen flowering shrubs and tree species for the wetlands and the surrounding subsoil drainage areas.

The scientists told us that anything green and which flowered would be a sensation for people living at the edge of the Sahara. Brainstorming with Rachid and the Sheikh, we decided to build the Wastewater Garden in the shape of the crescent moon, a symbol of some religious significance in the *Maghreb*.

Everybody, including the Algerian government, loved our green crescent moon Wastewater Garden, and it has performed superbly, in spite of the extreme heat, sometimes above 105 degrees Fahrenheit (41 degrees centigrade) and periodic sandstorms. We have recently joined forces with a French wetlands company to design a dozen more constructed wetlands, to serve towns of up to 10,000 people, all over Algeria.

Restoring Eden – A Project in the Cradle of Civilization

To go to Iraq quite frankly terrified me. My first trip was in 2011. I said goodbye to friends, with more than a passing thought that this might be a final farewell.

But the prospect of doing something meaningful in a place of profound significance proved irresistible. The more I learned about

Mesopotamia, a land whose history extends back into the mists of time, the more my interest deepened. What has happened there recently ranks as one of the greatest crimes against our biosphere.

In the south of Iraq there is one of world's largest natural wetlands, between the two great rivers, the Tigris and Euphrates, which come down from Turkey and flow through Iraq, draining into the Arabian Gulf. This is the "Fertile Crescent," where agriculture began. The present-day inhabitants of this delta region, known as Marsh Arabs, have evolved a unique way of life, creating small islands from wetland mud, using the reed plants to build their homes and their boats, living side by side with their water buffalo.

Marsh Arabs follow Shi'ite Islam, and when the Shi'ites rose up in rebellion against the government of Saddam Hussein, in the early 1990s, the dictator stopped at nothing to hunt down the rebels hiding in the marshlands. Saddam ordered the diversion of the Tigris and Euphrates Rivers, building canals that extended hundreds of kilometers, turning these historic and important wetlands to desert. With their way of life abruptly shattered, a half million or more Marsh Arabs, whose ancestors had lived there for tens of thousands of years, were forced to leave.

When Saddam's regime was overthrown, a national environmental organization, Nature Iraq, worked with the locals to literally punch holes—with tractors and shovels—in the diversion canals. The two mighty rivers again flowed along their original routes. Now, about half of the wetlands have recovered, and hundreds of thousands of Marsh Arabs have returned to resurrect a culture so intimately entwined with the wetlands.

This story of ecological devastation followed by restoration, along with the civil and religious strife that has devastated the region, inspired photographer Meridel Rubenstein to conceive of a symbolic project. She termed it Eden in Iraq: Ecological, Cultural Restoration

Mark with Jassim Al-Asadi, head of Al-Chibaish office of Nature Iraq, Meridel Rubenstein, the artist who conceived of the Eden in Iraq project, Davide Tocchetto, and the Mohammed Shrine family.

through Art, Design and Environmental Science for it is believed that, if indeed there is a physical location for the biblical garden, it was here in the Fertile Crescent. Meridel lives about three miles away from Synergia Ranch, in New Mexico. When she heard about my work with Wastewater Gardens, her symbolic art project became infused with new energy. "Wouldn't it be even more symbolic if the restoration of the Garden of Eden was done using wastewater?" By a strange synchronicity, Nature Iraq was also referring to its work in the area as *Eden Again*. We made contact with Azzam Alwash, an Iraqi-American engineer who founded Nature Iraq (N.I.), and Jassim Al-Asadi, a water engineer who heads N.I. in Al-Chibaish (and like Moses, was literally born in the marshes in a reed boat). We then began seeking support for a project among the Marsh Arabs.

There is no sewage treatment in the cities that have sprung up as Marsh Arabs have returned. At most, there is a network of pipes that pumps wastes to a "switch plant," which then sends raw sewage into the marshes or into the Euphrates, which is also the source of the drinking water.

We have developed plans for communities and towns to serve as demonstration sites for our "Restoration of Eden" Wastewater Gardens and Art project. Dr. Davide Tocchetto, an Italian agronomist who worked for several years with Wastewater Gardens International, has joined the team, to which he brings long experience with very large constructed wetland systems serving many thousands of people. The design team also includes Prof. Peer Satikh of the Nanyang Technological University (NTU), Singapore (where Meridel is a professor) and Prof. Sander van der Leeuw, Arizona State University. A grant from NTU has enabled our design work and several site visits. The project will not only improve people's health by treating wastewater and preventing contamination of the rivers and the wetlands, it may also set a standard for the whole country, some 80% of which lacks any sort of conventional sewage treatment system, even in large cities like Baghdad.

The challenges for the Marsh Arabs and their wetlands remain great. Because of upriver water diversion for cities and agriculture in Turkey and northern Iraq, water flow in the Tigris and Euphrates is less than half of what it used to be. Dams and diversions are being used to keep the wetlands healthy, but recently water flow out of the wetlands was stopped simply to prevent them drying out. Reduced water flow means increased salinity, which is bad for the reeds, the water buffalo, and the Marsh Arabs.

Recent victories, however, include the designation of the wetlands as Iraq's first protected area: the Mesopotamian National Park. Our first Wastewater Garden projects will be for two Marsh Arab towns of 15,000 to 45,000 inhabitants. We have received approval and backing from the Iraqi Ministries of Environment, Water Resources, and Protection of Iraqi Wetlands, and the governate of the Dhi Qar province, who see the importance of finding affordable and ecologically-sound solutions to sewage pollution problems.

On subsequent trips to Iraq I became enamored of the people I met. The Marsh Arabs are welcoming and profoundly gentle people. Their way of life, so deeply connected to the wetlands, is inspiring. They know their water buffalo well. A Marsh Arab can pick out one that belongs to him, from among thousands, because he literally lives with it. The sight of a water buffalo "commuting to work," swimming through the wetlands, moves the soul. Yet, how I laughed when I learned that Marsh Arabs, in their reed boats, navigating the maze of waterways twisting through the vast wetlands, also carry cell phones!

The author in the recovering southern Iraqi marshes (2011), navigated through the myriad of channels between the reeds by boatman Emad Attah Saleh Al-Asadi.

11
Preserving the Planet, One Flush at a Time!

"Man is the most extravagant accelerator of waste the world has ever endured ... and his besom of destruction in the uncontrolled hands of a generation has swept into the sea soil fertility which only centuries of life could accumulate."
—F. H. King, in *Farmers of Forty Centuries.*

WE LIVE IN A TIME OF TRANSITION. To change our ways is more challenging than I had first anticipated. My longtime friend and mentor, John Allen, (especially in matters philosophical and indeed, more importantly, applied philosophy) frames the issue in the following way:

In many of the great religions of the world, like Hinduism, there is a sacred trinity; in this case the deities, Brahma, Vishnu (of whom Krishna is a manifestation), and Shiva. Brahma is in charge of *creation*, Shiva in charge of *destruction* (getting rid of the old and creating a space for something new to arise), and Vishnu is the balancing force between the two, the maintenance man deity. It's no use creating if you can't keep maintaining it and keep it functioning.

Our modern way of life puts huge emphasis on creativity and manifestation, while the Shiva-like "counter-culture" wants to bring about an end to the old. Where's Vishnu in all this, maintaining equilibrium, keeping the show on the road? Our throw-away consumer society, with its planned obsolescence, is anathema to what nature does. Nature recycles everything. Ecosystems neither generate nor dispose of waste. It's this continuity and maintenance of life that astound us when we really contemplate the workings of the biosphere of which we humans are a critical part.

Maintenance, Main Tenets, Maintain Us

It's time to talk about maintenance, and the consequences of a lack thereof! I admit to having been naively optimistic about the potential for consciousness-raising in people involved in treating and recycling their shit, turning it into flowers and fruit, right under their noses. But I was so busy dwelling on the low-maintenance needs of constructed wetlands compared to high-tech sewage treatment systems that I failed to foresee that some people would apparently interpret that as meaning *no* maintenance.

A friend in Bali described it succinctly. If you have a high-tech system, you know that it's going to need maintenance, spare parts, the addition of chemicals, dealing with mechanical breakdowns. In short, you're prepared for it being a pain to keep working. With constructed wetlands, some act as though it all carries on by itself without needing any attention. So I'm forever expanding our operating and maintenance manual, which continues to lengthen as I come across truly bizarre things.

Who would have thought, for example, that anyone would pave over a septic tank so you can't get at it? In our systems, we put a filter[3] at the discharge end of the septic tank. We go to some lengths to ensure there are easily accessible airtight covers on both the inlet and discharge sides so that accumulation of sludge can be monitored. In a recent case, location not revealed to save embarrassment, a beautiful tiled floor had to be broken to rediscover exactly where the septic tank was! The same thing has occurred with a control box or two.

Then there was an urgent e-mail about an inlet pipe that was blocked in one of our systems in Indonesia. That had never happened before. When it was unplugged, it was found to be full of toilet paper. How could that be? Well, the owners had added another building with a bathroom, and connected its toilet directly to the constructed wetland, bypassing their septic tank entirely! I make passionate pleas that septic tanks be large enough to avoid problems and to minimize the need for pumping and maintenance. If there are problems in Phase 1 (the separation of solids), they are likely to affect the whole system. Little did I suspect someone might not enlarge an existing septic tank so as to accommodate new toilets or build another septic tank, but would instead send the wastewater directly to the wetland.

Then there is the issue of the plants and the aesthetics of a constructed wetland. We know plant biodiversity decreases as systems mature. But, like every garden, a constructed wetland needs its taller vegetation to be cut back or pruned so it doesn't shade out smaller plants. Surrounding vegetation, especially trees, can't be allowed to block sunlight reaching the Wastewater Garden or it won't work to full capacity. Yet, at a number of our Mexican systems this basic maintenance protocol was ignored.

At some of the hotels and condominiums where Wastewater Gardens have been installed, I have seen that while other parts of the grounds and landscape were kept immaculate, no one had tended to or pruned the plants in the constructed wetlands, possibly since they were first installed. Perhaps a certain shit phobia persists even when there's a recycling system and thus a reluctance to take care of the system. Happily, there are far more examples of people taking pride in their wetlands systems and maintaining them so that they continue to look lovely and perform superbly.

3. We used filters made by Zabel Industries in the US and by Taylex in Australia, which can be easily removed, washed with water, and put back.

"Cast Down Your Buckets Where You Are!"

As far back as 1911, F.H. King, the former head of the US Soil Conservation Service, made plain the coming crisis chemically-based agriculture was leading to:

> "When we reflect upon the depleted fertility of our own older farm lands, comparatively few of which have seen a century's service, and upon the enormous quantity of mineral fertilizers which are being applied annually to them in order to secure paying yields, it becomes evident that the time is here when profound consideration should be given to the practices ... maintained through many centuries, which permit it to be said of China that one-sixth of an acre of good land is ample for the maintenance of one person, and which are feeding an average of three people per acre of farmland in the three southernmost of the four main islands of Japan ... the people of the United States and of Europe are pouring into the sea, lakes, or rivers and into the underground waters [millions of] pounds of nitrogen... phosphorus and potassium, and this waste we esteem one of the great achievements of our civilization. In the Far East, for more than thirty centuries, these enormous wastes have been religiously saved ... Compelled to solve the problem of avoiding such wastes ... they 'cast down their buckets where they were.'" [4]

King is alluding to the delightful story told by the great African-American educator, Booker T. Washington: "A ship lost at sea for many days suddenly sighted a friendly vessel. From the mast of the unfortunate vessel was seen a signal, "Water, water; we die of thirst!" The answer from the friendly vessel came back at once: "Cast down your bucket where you are." And a second, third, and

4. F.H. King, *Farmers of Forty Centuries*, 1911, reprinted by Rodale Press, Emmaus, PA, pp 193-198.

fourth signal for water was answered, "Cast down your bucket where you are." The captain of the distressed vessel, at last heeding the injunction, cast down his bucket and it came up, full of fresh sparkling water from the mouth of the Amazon River." The point of this story is that our assumptions, addictions, and preconceptions blind us to the obvious.

The "Fecesphere"

"The Biosphere lives on the Necrosphere"
(the sphere of the dead and decaying).

I remember listening to Ramon Margalef, the brilliant Spanish ecologist, speak on the necrosphere at the first Closed Ecological Systems and Biospherics workshop, held at the Royal Society in London, in 1987. His talk changed how I thought about life. We are not only "standing on the shoulders of giants," as was said about our debt to the great minds who have preceded us. We, the living, are also standing upon and feeding off *all* the life that has come before us. Life literally feeds off what has gone before—all the organisms that have made the earth as it is, laying down and transforming materials, creating resources to further sustain and expand the world of the living.

Vladimir Vernadsky, the Russian scientist and thinker who pioneered modern understanding of the biosphere, began his career as a geologist and geochemist. After he discovered that many mineral deposits previously thought to be purely "geological" were in fact the "traces of bygone biospheres," he founded the science of biogeochemistry.

The oil, coal, and natural gas that fuel our industrial society are all formed from the remains of living organisms—plants and animals which have been transformed into hydrocarbons by burial in anaerobic wetlands. The ceaseless action of this *bio*-sphere, over the last four billion years, is why planet Earth is an oasis in space.

Vladimir I. Vernadsky (1863-1945) revolutionized scientists' understanding of the huge impacts humans have had on the planet, making intelligent cooperation with the biosphere essential for creating what he called the Noosphere.

What else does life live on? If you have read this far, I think you know the answer! Shit. All of life is interconnected. Life is like a planetary supermarket with all organic materials on sale, with buyers for each and every item. In this market, though, the buyers are also on sale. An "eat and be eaten" policy applies on Earth—and that does not only refer to the body. Every excretion, respiration, and other "byproduct" of living organisms is part of the complex food chain that sustains and increases life on this planet—a never-ending feast that keeps all nutrient cycles going.

There is a poignant Yiddish story in which a man assesses his life by all that he has consumed: the number of eggs, chickens, potatoes, onions, dill pickles, matzo balls, bottles of wine, glasses of water, breaths of air, you name it. What the story doesn't enumerate is the flipside. If all that goes in one end, it comes out the other.

The sphere of *shit* helps sustain the biosphere. We could call it the *coprosphere,* since coprolite is the word for *fossilized* shit, but use of obscure Latin or Greek words seems to sanitize the reality. Shit rarely becomes fossilized. Fresh, it's a valuable resource. What about the *manuresphere* or *septosphere*? My candidate is the *fecesphere.*

The quantity of shit produced in a year by animals exceeds their body weight. We animals consume a good portion of the plant biomass of the planet and then, by excreting unneeded residues, we put organic material and other essential elements back into circulation. The challenge we face is how to get these nutrients more effectively recycled, to sustain life, instead of polluting land and water and spreading disease.

We needn't worry about the contribution of other animal species. Wild animals shit in the ecosystems where they eat, supplying nutrients back to the soil. Animal populations hardly ever expand to a point where they cause serious pollution. "Do bears shit in the

Water buffalo walking home to their Marsh Arab families, southern Iraqi marshlands.

woods?" "Carrying capacity" is the ecological term for the limitations determined by the availability of resources in an ecosystem. Along with predators and disease, carrying capacities keep wild animal numbers at sustainable levels.

Animals in zoos, domestic pets, and livestock in pens and feedlots are a different matter. In these situations, the wastewater recycling solutions proposed in this book are equally viable. Constructed wetlands have worked well in agribusiness operations such as piggeries and dairies. Composting animal manure and returning it to farmland would be a major step toward recovering the humus and organic matter critical to soil fertility that has been lost in the past century due to the application of chemical fertilizers. It is a very distressing fact that the mechanized methods of present-day agribusiness cause the loss of five to ten tons of topsoil for every ton of grain produced, per year.

Zoos could compost their animal waste and sell it as exotic organic fertilizer. Constructed wetlands could add to the beauty and botanic diversity of zoos. People with domestic pets and a garden might consider composting the shit of their beloved ones. Is this

Opposite page: The Global Water Cycle showing how human actions (including agriculture and industry) interact with the water cycle. As an intrinsic part of the biospheric cycle, we must become aware of where the food and water that sustains us comes from, and where the wastewater we create goes.

too farfetched, especially for urbanites? There could be municipal pet doo-doo bins at the recycling center, next to the glass, plastic, and metal collection bins. Beats carrying it back to your apartment and putting it in the trash. There could even be cash incentives since urban compost generates revenue.

The fundamental obstacle to the proper functioning of this Fecesphere is us!

We humans are a global species and have a planetary impact unlike any other species that the biosphere has produced to date. Humans have been "liberated" from the limitations of the carrying capacity of local ecosystems. Think about trying to forage for food from the trees and grasses nearby, in your favorite city park or garden. How many people could be sustained in this way? Nowadays, humans rely on complex systems of food production and distribution. Further, we are relentlessly increasing our harvesting of the earth's resources with over-fishing, over-hunting wild animals, and destroying their habitat, converting evermore "wild land" to farms and rangeland. We produce additional animal protein through evermore intensive factory farming.

Those resources that we consume turn into metabolic "wastes" (shit and piss) and industrial waste that we spend more resources to detoxify and then dump, causing pollution of our waters and oceans. Since we don't return any of the nutrients in our bodily effluents to the land, the net effect is to further impoverish the soils that produced our food.

In sum, we are rapidly using up the earth's store of capital resources, such as fossil fuels and minerals, and we live in ever larger urban areas which have to be supplied with what we extract from the remaining open spaces.

Hmm. Doesn't sound good, does it? But wait, there's some light at the end of the tunnel! There's a way back from the brink.

THE GLOBAL WATER CYCLE

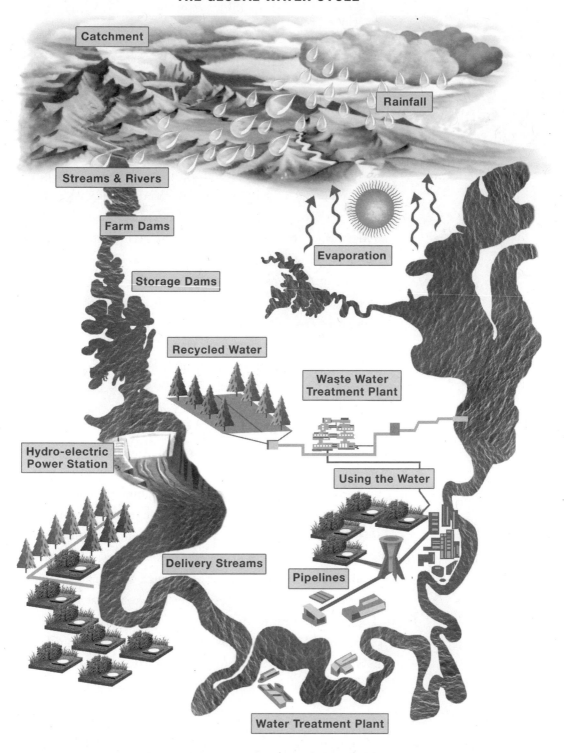

Getting There From Here

In recent years we have been deluged with information about sustainability, eco-this and eco-that. "Environmentally-friendly," "green" products line supermarket shelves. We humans are a thinking species, but undoubtedly we are neither the *only* thinking species on the planet nor the most intelligent, or we wouldn't be in this mess. We're certainly the only species with the technical know-how to have created a global culture and infrastructure. That's a good thing because how we'll get out of this mess is by both changing our habits and inventing appropriate technologies for dealing with our shit and our other problems.

I interpret all this sustainability talk as a healthy indication that we are becoming more and more, if somewhat uncomfortably, cognizant that things must change. "Things that can't go on forever, don't" is the aphorism credited to Herbert Stein, who was Chairman of the Council of Economic Advisers under President Richard Nixon.

We are all aware that the future can rarely be predicted by simply extrapolating from the present. Among many hundreds of crazy examples, there is the now ludicrous report dating from the early 1900s about the dire future awaiting New York City as its population increased and more people needed transportation. This meant the prospect of a skyrocketing increase in the number of horses and horse-drawn carriages, and it was foreseen that Manhattanites would shortly be buried neck-high in horse manure! Henry Ford's "horseless carriage" helped avert that calamity, although that same invention set in motion a world of complex environmental problems which dwarf the simple worry of being buried in heaps of horseshit. The future is for us to create. How do we get there from here?

The Restoration Project:
Creating the Earth We Want and Need

Since the Industrial Revolution in Europe at the beginning of the nineteenth century, a technology-dependent way of life has spread throughout the world. While it has been a time of exuberant optimism, it's also been one of greed, ignorance, and immaturity with respect to the power unleashed by having such technologies at our command. Clearly, there have been great advances in science and engineering, public health care, and communication. In general, life expectancy and the standard of living have been raised, year after year. Unintended side effects and widespread environmental degradation have also made this period the most ecologically destructive in all the time humans have inhabited the earth.

Until recently, little heed was taken of the impact of rapid population growth and these new technologies on natural ecosystems, the planet's soil, water, and atmosphere. Simultaneously, as scientific understanding has progressed, we are beginning to appreciate the workings of the biosphere—our life-support system. Our knowledge is incomplete and by no means is it assured that the human species has acquired the wisdom to act on the information we now have. It is an unprecedented challenge. We will have to regulate our behavior and set limits on our demands if we are to have a sustainable, viable future, one that we say we want and need.

Rather than dwell on the privations such changes in current behavior may bring as we wean ourselves from the culture of consumption and wastefulness, let's look at the benefits of the coming transformation. Protecting our environment is not merely an altruistic act. It is in our self-interest.

As my colleagues and I learned in Biosphere 2, the health of humans is intrinsically connected to the health of the earth's bio-

A Marsh Arab dwelling in southern Iraq on islands constructed of layers of reeds and marsh mud.

sphere. We will enjoy a truly higher standard of living when we live in beautiful, healthy ecosystems and have the deep satisfaction that comes from realization of our roles as important components of a vibrant and ever-evolving biosphere.

We have the tools and the knowledge to begin this restoration of the biosphere. Efforts are being made to protect unique areas of biodiversity, and those of ecological importance and awesome beauty. Wilderness areas, wildlife refuges, national parks, national forests, biosphere reserves, and marine conservation areas are all new realities. Ecological Restoration and Ecological Engineering are now recognized disciplines.

There is an enormous amount of other organic matter treated as residential "waste"– everything from kitchen food scraps, cut grass, leaves, and tree branches. This contributes at least a third to the "solid waste" that cities bury in landfills or have to incinerate. The movement to ban organic matter from disposal in this manner has begun, and a number of cities around the world are turning to municipal composting as a way of changing this "problem" into a benefit and a source of revenue.

Green belts and food-producing greenhouses are currently found around many Chinese cities and are traditional in a number of European countries as well. Such strategies could be put in place worldwide. In the West, interest in urban farming and "vertical farming" using high-rise structures is a very welcome sign. Greening our cities is a win-win situation at all levels—from producing fresh and delicious food to reducing transportation costs and making cities more pleasantly livable. Humans need to be scrupulous in returning shit to the land and, like the earthworms, become soil-makers. Why? Like earthworms, we are ultimately supported by the health of our soils.

This is not a "zero-sum game." Life increases in diversity and complexity as it transforms the energy Planet Earth receives. Giving back what we take out is the key to learning how to sustain life. "Living off the increase" is a longstanding motto of Institute of Ecotechnics projects around the world. Restoring and enriching damaged ecosystems increases their health, their biomass, and diversity, and so increases what the system sustainably produces.

My thirty years of experience in waste recycling and soil regeneration enterprises around the world leads me to some suggestions and conclusions. I see them as a rational way of managing the fecesphere. Some can be implemented by individuals, others will require the re-engineering of various support technologies and infrastructure. We built them so we can redesign them. It isn't rocket science.

1. Separate shit from the water cycle wherever possible. This can be done at the source, by means of composting toilets, or at the end, by sending wastewater through zero-discharge reuse and recycling systems.
2. Use water of the appropriate quality, according to need. That means clean, potable water should be prioritized for

drinking, cooking, and bathing. Irrigation water should be water of a lower quality—for example, unpurified groundwater or surface water or appropriately treated and managed graywater.

3. Conserve water by installing low-water-use appliances (washing machines, toilets, showers, etc.) and by irrigating using water-efficient methods such as drip irrigation or subsurface irrigation, lessening evapotranspiration.

4. Use wastewater to create green belts around cities and to landscape at the smaller septic tank house or small community scale.

5. Treat and reuse shit locally wherever possible. Centralized sewage treatment is very costly and makes more difficult the greening, recycling use of wastewater since much larger quantities of wastewater are now concentrated in one place. Wastewater recycling and redistribution should be decentralized, minimizing infrastructure and energy costs.

6. Do not mix industrial waste with residential waste. Detoxify the former before any possible recycling. Just as residential wastewater can be treated and recycled as close to its source as possible, so should industry be responsible for cleaning up and recycling its wastewater at its source. If industry had to pay for the downstream costs of wastewater pollution, there would be economic incentives to use less toxic products, to develop methods for detoxifying ones that have no substitutes, and to recycle rather than dispose of wastewater.

7. Send the sludge and compost made from human shit back to the land in an economical way. If necessary, add the real cost of returning shit to the land as a previously unaccounted cost of agriculture and of restoring and maintaining the health and productivity of our soils.

Every Step You Take, Every Shit You Make ...

I once heard a Buddhist meditation teacher say that a really good way of transforming the irritation we feel when we're stuck in traffic or waiting for a traffic light to change is to practice *mindfulness*, the transformation of negative emotions by means of meditation. The teacher contrasted this to the "California approach," where everything must be perfect: subtle lighting, soft cushions, a beautiful setting, and complete quiet. The problem with that approach is that real life, our life, does not nor will it ever take place in a "perfect" environment. One needs to find a way to still the mind and create a circle of self-awareness even when all hell is breaking loose around us.

I would like to add a fecesphere meditation. Each time you go to the toilet to take a dump, be *mindful* of what you are doing. "Where does my water come from?" "Where does my shit go?" Then perhaps, investigate, find out. You will be way more connected to reality by trying this simple meditation, and you'll come to understand how life on this planet is indeed sustained. Then ask: "How can I make this activity healthier for my local ecosystem and indeed the biosphere?" "How can I change the world?"

The answer is not in the glorious, perfected hereafter ("There'll be pie in the sky when you die") but right now, beginning with understanding the "travel itinerary" of your shit.

How do we change the world, help create the earth we want and need?

No action is trivial or unimportant.

We change the world one small step at a time, one flush at a time.

Recommended Further Reading

THE HISTORY OF SEWAGE AND TREATMENT SYSTEMS:

Del Porto, David. *The Composting Toilet System Book*, Concord, MA: Center for Ecological Pollution Prevention, 1999

Rockefeller, Abby A. **"Civilization & Sludge: Notes on the History of the Management of Human Excreta"** in *Current World Leaders,* Vol. 39 (1996), No. 6, pp 99-113

Laporte, Dominique. *The History of Shit,* Cambridge, MA: MIT Press, 2000

Jenkins, Joseph. *The Humanure Handbook,* White River Junction, VT: Chelsea Green Publ., 1999

Pickford, John. *Low-Cost Sanitation,* London: ITDG Publ., 1995

Lewin, Ralph A. *Merde: Excursions in Scientific, Cultural and Socio-Historical Coprology,* New York: Random House, 1999

Winblad, Uno *Sanitation without Water,* London: Macmillan Education, 1985

Kahn, Lloyd *Septic Tank Owner's Manual,* Bolinas, CA: Shelter Publ., 2000

Hart-Davis, Adam. *Thunder, Flush and Thomas Crapper, an Encyclopedia,* London: Michael O'Mara Books, 1997

Barlow, Ronald. *The Vanishing American Outhouse: A History of Country Plumbing,* El Cajon, CA: Windmill Publ., 1989

Fahn, Latee. *The Waste of Nations: The Economic Utilization of Human Waste in Agriculture,* Montclair, NJ: Allenheld, Osmun & Co., 1980

ORGANIC FARMING AND COMPOSTING:

King, F.H. *Farmers of Forty Centuries,* Emmaus, PA: Rodale Press, 1911

Pfeiffer, Ehrenfried. *Biodynamic Gardening and Farming,* Three Springs, NY: Anthroposophic Press, 1938

Howard, Albert. *An Agricultural Testament,* Oxford University Press, UK, 1943

National Research Council. *Alternative Agriculture,* Washington DC: National Academy Press, 1989

Rodale, J.I. *The Complete Book of Composting,* Emmaus, PA: Rodale Press, 2000

BIOSPHERE 2 AND CLOSED ECOLOGICAL SYSTEMS:

Shepelev, Yevgeny Y. **"Biological Life Support Systems,"** in Calvin, M. (ed.) *Foundations of Space Biology and Medicine* (3) (274-308), Moscow: Academy of Sciences and Washington DC, NASA, 1975

Nelson, Mark & Gerald A.Soffen (eds.) *Biological Life Support Technologies: Commercial Opportunities*. Oracle, AZ: Synergetic Press/Washington DC: NASA, 1990

Nelson, Mark. **"Bioregenerative Life Support for Space Habitation & Extended Planetary Missions"** in Churchill, S. (ed.) *Fundamentals of Space Life Sciences, Volume 2* (315-336). Malabar, FL: Orbit Books, 1997

Allen, John. *Biosphere 2: The Human Experiment,* New York: Penguin Books, 1991

H.T. Odum, **"Biosphere 2 Research: Past and Future"** in *Ecological Engineering Special Issue, Volume 13* (1-4) Amsterdam: Elsevier Science, 1999

Terskov, I.A. **"Closed System: Man-Higher Plants (Four Month Experiment)"** in *NASA TM 76452*, trans. Nauka Press, Novosibirsk (report on most advanced Russian work) Washington DC: NASA, 1981

Silverstone, Sally. ***Eating In: From the Field to the Kitchen in Biosphere 2,*** Oracle, AZ: Biosphere Press, 1993

Folsome, Clair E. **"The Emergence of Materially Closed System Ecology"** in Polunin, N. (ed.) *Ecosystem Theory and Application* (269-288), New York: John Wiley & Sons, 1986

Alling, Abigail, Mark Nelson & Sally Silverstone. ***Life under Glass: The Inside Story of Biosphere 2,*** Oracle, AZ: Biosphere Press, 1993

Kelly, Kevin. ***Out of Control: The New Biology of Machines, Economic and Social Systems*** New York: Addison-Wesley Publ., 1994

Allen, John and Mark Nelson. **Space Biospheres,** Oracle, AZ: Synergetic Press, 1985

ECOLOGICAL ENGINEERING, SYSTEMS ECOLOGY AND THE WORK OF H.T. AND E.P. ODUM:

Odum, H.T. **"Ecological Engineering and Self-Organization"** in Mitsch, W. & S. Jorgensen,. (eds.) *Ecological Engineering: An introduction to Ecotechnology* (79-101), New York: Wiley & Sons, 1991

Odum, E.P. *Ecology.* Philadelphia: Holt, Reinhart & Winston, 1966

Odum, H.T. *Ecological and General Systems: An Introduction to Systems Ecology,* Niwot, CO: University of Colorado Press, 1994

Odum, E.P. **Ecology and our Endangered Life Support System,** Sunderland, MA: Sinauer Associates Publ., 1993

Odum, H.T. **Environmental Accounting: Emergy and Decision Making.**
New York: John Wiley, 1996

Odum, H.T. **"Scales of Ecological Engineering"** in *Ecological Engineering,* Vol. 6, (1996) (7-19)

CONSTRUCTED WETLANDS AND WASTEWATER GARDENS

Wolverton, Billy C. **"Aquatic Plants And Wastewater Treatment (An Overview)"** in Reddy, K.R., Smith W.H. (eds.) *Aquatic Plants for Water Treatment and Resource Recovery* (3-15) Orlando, FL: Magnolia Publ., 1987

Nelson, Mark, M. Finn, et al. **"Bioregenerative Recycle of Wastewater in Biosphere 2 Using a Created Wetland: Two Year Results,"** *Journal of Ecological Engineering,* (13) (1999): (189-197)

Moshiri, G.A. (ed.) ***Constructed Wetlands for Water Quality Improvement,*** Boca Raton, FL: Lewis Publ., 1992

United States Environmental Protection Agency. ***Constructed Wetlands Treatment of Municipal Wastewaters* (EPA/625/R-99/010)** Cincinnati: Office of Research & Development, Cincinnati, Ohio, 2000

Steiner, G.R. & J.T. Watson. ***General Design, Construction and Operation Guidelines: Constructed Wetland Wastewater Treatment Systems for Small Users, including Individual Residences,*** Chattanooga, TN: Tennessee Valley Authority, 1993

Nelson, Mark. **Limestone Mesocosm for Recycling Saline Wastewater in Coastal Yucatan, Mexico** (PhD dissertation), Gainesville, FL: University of Florida, Dept. of Environmental Engineering Sciences, 1998

Nelson, Mark, H.T. Odum, et al. (2000) **"Living off the Land: Resource Efficiency of Wetland Wastewater Treatment"** in *Advances in Space Research* (27) (9): (1546-1556)

Reed, S.C. & R.W. Crites. *Natural Systems for Waste Management and Treatment,* New York: McGraw-Hill, 1996

United States Environmental Protection Agency. *Subsurface Flow Constructed Wetlands for Wastewater* Treatment (EPA/832/R-93/008), Washington DC: EPA Office of Water, 1993

Kadlec, R.H. & R. L. Knight, *Treatment Wetlands,* Boca Raton, FL: Lewis Publ., 1996

Nelson, Mark. **"Wetland Systems for Bioregenerative Reclamation of Wastewater: From Closed Systems to Developing Countries"** in *Journal of Life Support and Biosphere Science* (Vol.5, 1998) (3): (357-369)

GRAYWATER RECYCLING

Ludwig, Art. *Create an Oasis with Greywater,* Santa Barbara, CA: Oasis Design, 2000

Ludwig, Art, *Branched Drain Greywater Systems,* Santa Barbara, CA: Oasis Design, 2000

OTHER BOOKS OF INTEREST

Allen, John P. *Me and the Biospheres: A Memoir by the Inventor of Biosphere 2*, Santa Fe, NM: Synergetic Press, 2009

Meyer, Kathleen *How to Shit in the Woods,* Berkeley, CA: Ten Speed Press, 1989

Mitsch, W.J. *Wetlands,* New York: Van Nostrand Rheinhold, 1993

Lapo, Andrei. *Traces of Bygone Biospheres.* Moscow: Mir Publ., 1987

Vernadsky, V.I. *The Biosphere,* Oracle, AZ: Synergetic Press, 1986

Darwin, Charles *The Formation of Vegetable Mould, through the Action of Worms,* New York: Appleton & Co., 1882

Links

Wastewater Gardens International www.wastewatergardens.com
Institute of Ecotechnics www.ecotechnics.edu
Biosphere 2 Project www.biospherics.org
Synergia Ranch, New Mexico www.synergiaranch.com
Birdwood Downs Station, W. Australia www.birdwooddowns.com
IDEP Foundation, Bali, Indonesia www.idepfoundation.org
Global Ecotechnics Corporation www.globalecotechnics.com
Biosphere Foundation www.biospherefoundation.org
Las Casas de la Selva www.eyeontherainforest.org
Nature Iraq www.natureiraq.org
Eden in Iraq Project www.meridelrubenstein.com/eden-in-iraq

Credits

Most of the photos are from my own scrapbook, but I owe special thanks for some of the beautiful photos of Wastewater Garden systems provided by Emerald Starr, Biosphere Foundation, in Bali, Gonzalo Arcila in Mexico, Thrity Vakil in Puerto Rico, Hans Leenaarts in Australia, and Florence Cattin in Indonesia and Algeria, and Meridel Rubinstein in Iraq. To Deborah Parrish Snyder, Marie Harding and Robert Hahn for various photos illustrating the history of our projects. And to Gill Kenny and Abigail Alling for photos from the Biosphere 2 project. If I've missed anyone's photo credit, let me know.

Other credits due to:

p. 7 – The Royal Loo © Audrey Bergner
p. 8, 9 – Valveless Water Preventer, Thomas Crapper Advertisements © Thomas Crapper & Co., Ltd.
p. 39 – Photo by Abigail Alling, the Biosphere Foundation
p. 50 – Permission of the *Arizona Daily Star,* Tucson
p. 52 – Two-section septic tank © ASP Environmental
p. 53 – Bed system © National Environmental Services Center at West Virginia University
p. 58 – Modern sewage plant © Kekyalyaynen
p. 72 – Cypress Swamp © visitsouth.com
p. 111 – Artist unknown

I wish to acknowledge my colleagues at the Planetary Coral Reef Foundation (PCRF), through which we built many of the systems mentioned in Mexico, Indonesia, and Australia. PCRF was founded in 1991 to address the worldwide demise of coral reefs through scientific research, protection and education. In its efforts to fight water pollution, PCRF significantly contributed to the creation of Wastewater Gardens International and its network. It continues to seek to further extend this ecotechnology in highly sensitive areas, in particular in countries where acquatic or marine ecosystems are severely endangered. PCRF is a division of the Biosphere Foundation (www.biospherefoundation.org).

Wastewater Gardens International Representatives

FOUNDER, PRINCIPAL DESIGNER:

Dr. Mark Nelson / Wastewater Gardens International
1 Bluebird Court Santa Fe - NM 87508 - USA
Email: nelson@biospheres.com
website: http://www.wastewatergardens.com

REGIONAL DIRECTOR AND INTERNATIONAL LIAISON,
DESIGN & IMPLEMENTATION OFFICER:

Florence Cattin / Wastewater Gardens International
Aptdo Correo N° 54 - 11150 Vejer de la Frontera, Cádiz - Spain
Email: fc@internationalsolutions.org - Alternate email: fcattin@gmail.com,
website: www.wastewatergardens.com

SOUTHEAST ASIA / INDONESIA REPRESENTATIVE

I Gede Sugiartha / Yayasan IDEP Foundation
Br. Bucuan - Desa Batuan - Sukawati - Gianyar 80582 - Bali – Indonesia
Email: gede@idepfoundation.org, info@idepfoundation.org
website: http://www.idepfoundation.org

EUROPE REGIONAL AND INTERNATIONAL
REPRESENTATIVE, AFFILIATE:

Dr. Davide Tocchetto
Via C. Maltoni, 25 - 31044 Montebelluna - Italy
Email: davide.tocchetto@libero.it

MEXICO, REGIONAL AND INTERNATIONAL REPRESENTATIVE:

Gonzalo Arcila & Ingrid Datica
Villa Balamek. Lote 27 Secc. "F" Akumal Norte, Tulum. Q. Roo CP
77729 MEXICO.
Email: gonzalo@akumal.com

EUROPE CONTACT / REGIONAL DIRECTOR FOR POLAND:

Dr. Andrzej Czech / Ecofrontiers, Natural Systems
Uherce Mineralne 285, 38-623 Uherce Mineralne - Poland
Email: czech@naturalsystems.pl; ac@naturalsystems.pl
website: www.naturalsystems.pl

AUSTRALIA CONTACT:

Hans Leenaarts, General Manager / Birdwood Downs Company
Gibb River Rd./ PO Box 124 - Derby - West Australia 6728
Email: info@birdwooddowns.com
website: www.birdwooddowns.com

WEST AUSTRALIA REGIONAL REPRESENTATIVE:

Andrew Hemsley / Integrated Natural Systems
9 High Street, Busselton, WA Australia 6280
email: andrew@integratednaturalsystems.com.au
website: www.integratednaturalsystems.com.au

BELIZE REPRESENTATIVE:

Lucien Chung MSc. PEng. / Chung's Engineering Co. Ltd.
8 Dickenson St. - Belize City, Belize.
Email: lucienchung@gmail.com

INDIA LIAISON:

Malini Rajendran / MIECOFT - Mission to Implement Eco-Friendly Technology
110, Shikha Apartments, 48, I.P.Extension - Delhi 110092 - India
Email: miecoft@gmail.com

Index

U

urban farming, 191

V

vertical farming, 191

W

Wastewater Gardens International, 145, 170, 176
water buffalo, 175, 177
water cycle, 33, 186, 191
water hyacinth, 37, 43, 153
water lotus, 156, 159
Western Australia, 118, 122, 134, 137, 140-141, 145
World Bank, 153
World Health Organization, 152

Z

zero discharge, 60, 81
zero-sum game, 191

About the Author

Mark Nelson, PhD, is an ecosystem engineer and researcher, and one of the original "Biospherians." He is Chairman, CEO, and a founding director of the Institute of Ecotechnics, a UK and US nonprofit organization consulting on several demonstration projects working in challenging biomes around the world. He is head of the Biospheric Design Division, Global Ecotechnics Corporation. Founder and director of Wastewater Gardens International, he designs and implements sewage treatment and recycling systems using constructed wetlands.

Dr. Nelson was a member of the eight-person crew inside Biosphere 2, the 3.15 acre materially-closed facility near Tucson, Arizona, during the first two-year closure experiment from 1991-1993. He has worked for several decades in closed ecological system research, ecological engineering, the restoration of damaged ecosystems, desert agriculture and orchards and wastewater recycling.

He holds a PhD in Environmental Engineering Sciences from the University of Florida; an M.S. from the School of Renewable Natural Resources, University of Arizona; and a B.A. in Philosophy and Biological Sciences from Dartmouth College.

Mark's Wastewater Gardens projects have taken him around the world, including the coast of Yucatan, Mexico; the high desert grassland in New Mexico; the semi-arid tropical savannah of West Australia; the lush tropics of Bali and Southeast Asia; and most recently, the southern marshlands of Iraq.

www.synergeticpress.com